T0234701

SpringerBriefs in Applied Sciences and Technology

Computational Intelligence

Series Editor

Janusz Kacprzyk, Systems Research Institute, Polish Academy of Sciences, Warsaw, Poland

SpringerBriefs in Computational Intelligence are a series of slim high-quality publications encompassing the entire spectrum of Computational Intelligence. Featuring compact volumes of 50 to 125 pages (approximately 20,000-45,000 words), Briefs are shorter than a conventional book but longer than a journal article. Thus Briefs serve as timely, concise tools for students, researchers, and professionals.

More information about this subseries at http://www.springer.com/series/10618

Tomé Almeida Borges · Rui Neves

Financial Data Resampling for Machine Learning Based Trading

Application to Cryptocurrency Markets

 Springer

Tomé Almeida Borges
Instituto Superior Técnico
Instituto de Telecomunicações
Lisbon, Portugal

Rui Neves🆔
Instituto Superior Técnico
Instituto de Telecomunicações
Lisbon, Portugal

ISSN 2191-530X   ISSN 2191-5318   (electronic)
SpringerBriefs in Applied Sciences and Technology
ISSN 2625-3704   ISSN 2625-3712   (electronic)
SpringerBriefs in Computational Intelligence
ISBN 978-3-030-68378-8   ISBN 978-3-030-68379-5   (eBook)
https://doi.org/10.1007/978-3-030-68379-5

© The Author(s), under exclusive license to Springer Nature Switzerland AG 2021
This work is subject to copyright. All rights are solely and exclusively licensed by the Publisher, whether
the whole or part of the material is concerned, specifically the rights of translation, reprinting, reuse of
illustrations, recitation, broadcasting, reproduction on microfilms or in any other physical way, and
transmission or information storage and retrieval, electronic adaptation, computer software, or by similar
or dissimilar methodology now known or hereafter developed.
The use of general descriptive names, registered names, trademarks, service marks, etc. in this
publication does not imply, even in the absence of a specific statement, that such names are exempt from
the relevant protective laws and regulations and therefore free for general use.
The publisher, the authors and the editors are safe to assume that the advice and information in this
book are believed to be true and accurate at the date of publication. Neither the publisher nor the
authors or the editors give a warranty, expressed or implied, with respect to the material contained
herein or for any errors or omissions that may have been made. The publisher remains neutral with regard
to jurisdictional claims in published maps and institutional affiliations.

This Springer imprint is published by the registered company Springer Nature Switzerland AG
The registered company address is: Gewerbestrasse 11, 6330 Cham, Switzerland

Tomé Almeida Borges

To my Grandparents and Beatriz

Rui Neves

To Susana and Tiago

Preface

A financial market describes any marketplace where buyers and sellers participate in the trade of assets such as currencies, equities, bonds and derivatives. Financial activities play an important role in the world economy and influence worldwide economic development. On this account, financial time series forecasting has attracted increasing attention among traders and researchers in the fields of finance, engineering and mathematics for its theoretical possibilities and practical applications.

Financial time series forecasting is a challenging task. This is primarily because of the uncertainties involved in the movements of markets. Many diversified factors cause an impact in financial markets such as political events, general economic conditions and trader's expectations. Therefore, financial time series are characterized by non-stationarity, heteroscedasticity, discontinuities, outliers and high-frequency multi-polynomial components making the prediction of market movements quite complex.

The complex characteristics of financial time series and the immense volumes of data that must be analyzed to successfully accomplish the task of forecasting financial time series have driven the adoption of more sophisticated methods, models and simulation techniques. Lately, machine learning or data mining techniques, widely applied in forecasting financial markets, have been offering much better results than simple technical or fundamental analysis static strategies. Machine learning methodologies are able of uncovering patterns and predict future trends in order to identify the best entry and exit points in a financial time series with the intention of achieving the highest returns with the lowest risk.

As a result of the tremendous growth and remarkably high market capitalizations reached, cryptocurrency are now emerging as a new financial instrument and their popularity has recently skyrocketed. As of now, thousands of cryptocurrencies, each with unique objectives and characteristics, are being actively traded against other cryptocurrencies or fiat money. The cryptocurrency exchange market is this work's financial market of focus. These fast-paced markets, known for their volatility and intense swings were selected with the intent of testing and clearly proving the capabilities of this book's ensemble of machine learning algorithms based voting system, and most importantly, the resampling methods applied and described in this book.

The cryptocurrency market is relatively recent, thus, literature regarding machine learning forecasting and investment in the cryptocurrency exchange market is still uncommon, save for specifically Bitcoin. This current work, instead of focusing solely on Bitcoin, aims to increase the scope to a broader set of cryptocurrencies in order to measure and compare the performance of several machine learning methods. In this work, as is typically posed, the forecasting problem is defined as direction-predicting, where the objective is predicting whether a given cryptocurrency pair exchange price is increasing or decreasing while utilizing only technical analysis as input data.

Each of the learning algorithms in the proposed system achieved better results than the Buy and Hold (B&H) strategy for many of the 100 cryptocurrencies tested. The best results are achieved for the unweighted average with accuracies that reach 59.26% for time resampled series. Last but foremost, it follows that the system developed in this work is capable of generating significantly greater returns when utilizing any of the three alternative resampling methods rather than time resampled data as is standard.

Chapter 1 describes the problematic addressed by this book, the main goals and the document's structure. Chapter 2 contains the theory and key concepts behind the developed work, an introduction to the cryptocurrency technology and some notions regarding market analysis and forecasting. Secondly, an introduction to each of the learning methods employed in this work as well as some existing strategies that seek to address analogous problems are described. Chapter 3 documents the entire proposed system architecture in detail. Chapter 4 describes the data, metrics and the benchmark strategy used to perform system validation ending with the obtained results displayed. Chapter 5 summarizes this work, presents the respective conclusion and presents some suggestions for future work.

Lisbon, Portugal   Tomé Almeida Borges
August 2020   Rui Neves

Contents

Acronyms

Investment Related

%D	Fast D Stochastic Oscillator Component
%K	Fast K Stochastic Oscillator Component
ATR	Average True Range
B&H	Buy and Hold
CCI	Commodity Channel Index
CDS	Credit Default Swaps
EMA	Exponential Moving Average
MACD	Moving Average Convergence Divergence
MDD	Maximum Drawdown
OBV	On Balance Volume
OHLC	Open-High-Low-Close
ROC	Rate of Change
ROI	Return on Investment
RSI	Relative Strength Index
SMA	Simple Moving Average
TI	Technical Indicator
TR	True Range
USD	United States Dollar

Optimization and Computer Engineering Related

ANN	Artificial Neural Network
API	Application Programming Interface
ARIMA	Autoregressive Integrated Moving Average
CART	Classification and Regresstion Tree
EBNN	Elman Backpropagation Neural Network

EV	Ensemble Voting
GA	Genetic Algorithm
LDA	Linear Discriminant Analysis
LR	Logistic Regression
LSTM	Long Short Term Memory
MLP	Multi-Layer Perception
NLL	Negative Log-Loss
RF	Random Forest
RMSE	Root Mean Squared Error
RNN	Recurrent Neural Network
SVC	Support Vector Classifier
SVM	Support Vector Machine
XGB	Extreme Gradient Boosting

List of Figures

List of Tables

Chapter 1
Introduction

1.1 Work's Purpose

The main purpose of this book is to demonstrate the advantage of utilizing resampled financial data when forecasting the future movements of economic markets. With this intention, a realistic system capable of predicting the optimal market entry and exit points in cryptocurrency exchange markets was created through the maximization of negative logarithmic loss in order to obtain the maximum returns while minimizing the risk. This system obviously had to be apt to be implemented in a real life investment scenario. The optimal market entry and exit points are derived from predicting the direction of price variations for several cryptocurrency pairs with the smallest forecasting error leading to the reduction of investment risk.

To accomplish this objective, firstly four different data resampling processes are computed to be later compared. Secondly, four different machine learning methods are employed to predict the future price directions. In the end, the predictions from each method are aggregated into a fifth prediction method as an attempt to obtain a better performance relatively to each individual machine learning method.

Finally, each of the five predictions for each of the four resampling methods (20 instances per market in total), are simulated through backtest trading in a real case scenario. The economic results as well as other metrics associated with machine learning forecasting obtained from the simulation of the several forecasting methods are benchmarked against a simple Buy and Hold strategy and compared among each other.

© The Author(s), under exclusive license to Springer Nature Switzerland AG 2021
T. A. Borges and R. Neves, *Financial Data Resampling for Machine Learning Based Trading*, SpringerBriefs in Computational Intelligence,
https://doi.org/10.1007/978-3-030-68379-5_1

1.2 Main Contributions

The main contributions of this work are:

- Development of a framework consisting of several supervised machine learning procedures to trade in a relatively new market, the Cryptocurrencies Market;
- Compare the performance of 5 different types of forecasting trading signals amongst themselves and with a Buy and Hold strategy as baseline;
- Compare the difference in returns and overall results between a financial time series resampled according to a parameter derived from trading activity, namely, a logarithmic, a fixed or a percentual variation to a commonly used time sampled time series, as baseline.

1.3 Document Structure

The structure for this book is the following:

- Chapter 2 contains the theory and key concepts behind the developed work. It begins with an introduction to Cryptocurrencies followed by some notions regarding market analysis and forecasting. An introduction to each of the learning methods employed in this work is given as well. Finally, some existing strategies that seek to address analogous problems are described.
- Chapter 3 documents the entire proposed system architecture in detail, explaining each of the main modules that constitute it.
- Chapter 4 describes the data, metrics and the benchmark strategy used to perform system validation. Afterwards, the obtained results are displayed.
- Chapter 5 summarizes this work, supplies the respective conclusion and presents some suggestions for future work.

Chapter 2
Background and State-of-the-Art

2.1 Cryptocurrencies

Nowadays, payments may be sent directly from one party to another through a peer-to-peer version of electronic cash that does not require going through a financial institution. The usage of these digital currencies allows for more innovative ways of payment and financing goods and services. One digital currency, Bitcoin, stood out among the rest. Bitcoin is considered to be the first of the newest subset of digital currencies, the *cryptocurrencies*.

The path to creating the first cryptocurrency as they are known today is paved by failed attempts at creating digital currencies. The earliest ideas of applying cryptography to currency came from David Chaum in 1983 [1], where he figured out how to avoid double-spending[1] while keeping the system anonymous through blind signatures.

Since the start of digital currencies with David Chaum, many other versions of digital currencies have been launched and failed until the years of 2007 and 2008, when a severe financial crisis rapidly developed and spread internationally into a global economic shock, resulting in several bank failures as well as the near collapse of credit markets [2] causing fear in the failure of government-controlled agencies and therefore generating interest in alternatives [3]. As OECD's (Organization for Economic Co-Operation and Development) Secretary General Ángel Gurría said in 2009 [4]:

> The global financial and economic crisis has done a lot of harm to the public trust in the institutions, the principles and the concept itself of the market economy.

The 2007–2008 financial crisis showed how unsafe it was to empower trusted third parties and how protective the government was of the banking system. This crisis imposed serious costs to the governments that bailed out overextended borrowers and/or lenders. These bailouts lead to higher fiscal deficits and public debt, banks no

---

[1] Flaw in which the specific same amount of coins is spent more than once.

© The Author(s), under exclusive license to Springer Nature Switzerland AG 2021
T. A. Borges and R. Neves, *Financial Data Resampling for Machine Learning Based Trading*, SpringerBriefs in Computational Intelligence, https://doi.org/10.1007/978-3-030-68379-5_2

longer trusted each other and were hoarding significant liquidity as a precautionary buffer [5].

In the year 2008, in a white paper called: "Bitcoin: A Peer-to-Peer Electronic Cash System" [6], self published by the pseudonymous Satoshi Nakamoto, Bitcoin was first described. Bitcoin is quite remarkable as it put an end to 20 years of previously unsuccessful digital currencies. Bitcoin is a peer-to-peer version of electronic cash that allows online payments to be sent directly from one party to another without going through financial institutions [6]. It is the first digital currency to introduce the concept of decentralized cryptocurrency.

During the financial crisis, Bitcoin and its succeeding cryptocurrencies would benefit from these negative reactions aimed towards the banking system, whereas, similarly to commodities such as gold, these currencies offered an alternative for those whose faith in the fiat currencies was damaged. But most importantly, cryptocurrencies were thought to possess the positive characteristics of a backed currency without the perceived drawbacks of fiat currencies, namely, being under the control of third parties. Cryptocurrencies instead of relying on a single intermediary, such as a bank or credit card network, rely upon a large number of competing "miners" to verify transactions. Satoshi Nakamoto envisioned an online system that resembled one person handing cash to another, while preserving the services a third party is expected to perform [6]. To verify if each newly made transaction is valid, miners must confirm a set of rules outlined in the source code. The supply of cryptocurrency may or may not be limited, but the creation of coins follow a rule that no government or institution can change, avoiding the possibility of harmful inflationary policies [3].

Bitcoin and the remaining cryptocurrencies filled in an important niche by providing a virtual currency decentralized system without any trusted parties and without pre-assumed identities among the participants that supports user-to-user transactions [7].

Conversely to cash, coins or stocks, bitcoins aren't discrete and aren't stored. Owning a bitcoin simply means knowing a private key which is able to create a signature that redeems certain outputs. As long as owners keep their private keys hidden, this system prevents theft or the usage of bitcoins one does not rightfully own. A bitcoin is a record of a transaction between different addresses in the Bitcoin network [6]. This principle was fundamentally followed by the remaining cryptocurrencies.

In cryptocurrencies a transaction is a transfer of funds between two different addresses, also known as digital wallets. In simple terms, a transaction communicates to the network that a rightful owner of cryptocurrency units has authorized the transfer of some of those units to another owner. After this, the cycle may repeat itself; the new owner is now able to make new transfers with the received units, hence creating a chain of ownership.

To illustrate this, looking at Bitcoin's case, each transaction record contains one or more input, where an amount of bitcoins will be debited, and one or more output on the other side of the transaction, where bitcoins are to be credited. Transactions are like lines in a double-entry bookkeeping ledger [8]. The debited and credited amount of bitcoins may not add up to the same amount. Instead, outputs add up to slightly less than inputs and the difference represents an implied "transaction fee", a small payment collected by the miner node, the one responsible for including the transaction in the public ledger. The miners, through competitive computation are able to achieve a consensual blockchain, the authoritative ledger of all transactions, without requiring a central authority [8].

The transaction also contains proof of ownership for each transferred amount of bitcoins, in the form of a digital signature from the owner, which can be independently validated by anyone. In Bitcoin terms, "spending" is signing a transaction which transfers value from a previous transaction over to a new owner identified by a bitcoin address.

As previously stated, Bitcoin is known as the "original" cryptocurrency, it is awarded the invention of the blockchain technology. The term "altcoin" describes the hundreds of cryptocurrencies that preceded and are alternative to Bitcoin. Some altcoins just alternate some parameters or add some form of innovation to Bitcoin's three main technical components: transactions, consensus protocol and communication protocol. Nevertheless, these altcoins contain a completely separate blockchain and network. Each of these forks attempts, in its own way, to fix shortcomings or improve on Bitcoin.

Other altcoins, on the other hand, may not attempt to be currencies at all. Cryptocurrencies may possess countless different purposes, such as social networks (e.g. Steemit), cloud computing (e.g. Golem) or data storage (e.g. Siacoin), decentralized trading exchanges (e.g. Binance Coin), powering the internet of things (e.g. IOTA) and the list goes on [8, 9]. The vast majority of cryptocurrencies are created with a specific aim that addresses different problems or aspects.

Both Bitcoin and Altcoins can be purchased, sold and exchanged for other currencies at specialized currency exchanges. These cryptocurrency exchanges, operate as a matching platform. That is, users do not trade with the exchange. Rather, they announce limit orders to buy and sell, and the exchange matches buyers and sellers when conditions of both the buyer and the seller are met. On the cryptocurrency trading market, much like the underlying principle where the Foreign Exchange market was built on, traders are essentially exchanging a cryptocurrency for another cryptocurrency or fiat currency.

Cryptocurrencies are commonly considered inadequate as a day-to-day currency due to the large fluctuations in price [10]. For instance, despite Bitcoin being initially created as a means of payment and actually being accepted in some stores and platforms, it cannot yet be considered a common medium of exchange [11]. Its argued it behaves as a speculative asset rather than a currency [12], or at least somewhere in between the two [13]. Even though many cryptocurrencies were created as a network to pay for goods or services, users commonly trade these as speculative investments, particularly Bitcoin [14].

In fact, the average monthly volatility in returns on Bitcoin is higher than in gold or a set of foreign currencies in US dollars, but the lowest monthly volatilities for Bitcoin are less than the highest monthly volatilities for gold and the foreign currencies [15].

Volatility is a reflection of the degree to which price moves. A stock with a price that fluctuates wildly and moves unpredictably is considered volatile, while a stock that maintains a relatively stable price has low volatility. A highly volatile market is inherently riskier, but that risk cuts both ways: high volatility may either generate larger losses or higher earnings.

The remarkable volatility visible in cryptocurrencies, among many reasons, is due to a lack of governance mechanisms and regulatory frameworks. Cryptocurrencies are a barely regulated space that exist on a collective user consensus of value and, to an extent driven by holders of large quantities of the total outstanding float of the currency [10]. It is not clear how investors with current holdings in the millions of US dollars, would liquidate a position that large into legal tender without severely moving the market.

The present volatility in cryptocurrencies fuels speculation which in turn leads to speculative bubbles in cryptocurrency markets and further volatility, hence creating a positive feedback loop [10]. Some significant price changes resemble a traditional speculative bubble [16, 17], which may occur when optimistic media coverage attracts investors. This obviously complicates an accurate assessment of any cryptocurrency's actual value as a currency and prompts merchants to exchange their cryptocurrencies for fiat currency quickly after receiving them as payment [18]. In fact, to protect retailers from these fluctuations, cryptocurrency payment service providers accept a set of currencies (typically the ones with largest market cap, such as Bitcoin, Ethereum and Litecoin) on behalf of the end receiver and the equivalent amount in legal tender is transferred at the current market exchange rate instead. This way sellers may either choose to receive payments in a combination of cryptocurrencies and legal tender, or avoid completely the risk of holding cryptocurrencies, with a small exchange fee charged by these payment service providers. The fee (which diminishes with volume) is usually around 1% on top of the additional fees imposed by banks or payout method.

Notwithstanding, both functions, asset and currency, arguably matter in the cryptocurrency market. If the main driver of demand was currency adoption, network effects[2] would be dominant and winner-take-all dynamics would take over. The lack of this dynamic in the recent period indicates that the financial asset function becomes more prominent. As Bitcoin's price and volatility increases, we see a substitution effect increasing the demand for other cryptocurrencies. Thus, the prices of all cryptocurrencies tend to move jointly [19].

In any case, in terms of number of new currencies, consumer base and transaction frequency, the market for cryptocurrencies has grown massively since 2008. Taking into account that no previous digital currency model had gotten even close to achieving these rates of success, this development is extraordinary.

---

[2]Positive effect that an additional user of a good or service has on the value of a specific product to others. Simply put, "the more users the merrier".

2.2 Cryptocurrency Exchange Market Analysis

The cryptocurrency exchange market, similarly to the Foreign Exchange Market (FOREX) market, allows trading among different cryptocurrencies or fiat currencies. There is a multitude of different trading platforms where these trades can be performed.

To study the change of rates in multiple cryptocurrency pairs, a time series can be built from sampling the market at a fixed time rate using historical data from a specific cryptocurrency exchange platform. Using a simple application programming interface (API) all historic data used in this thesis was retrieved from one single exchange: Binance. Binance is an exchange platform that implements a continuous double auction with an open order book and as of now, has the most traded volume according to [20].

Market trend movements can be characterized as bullish, bearish or sideways. A bull market occurs when prices are rising (uptrend), being characterized by optimism, investor confidence and expectations that strong results shall continue. The bear market is the opposite of a bull market, being characterized by falling prices (downtrend) and typically shrouded in pessimism. Lastly, when there is neither an uptrend or downtrend and the prices oscillate between a relatively narrow range, the market is said to be sideways.

In market trading, the terms *long* and *short* respectively refer to whether a trade was initiated by buying or selling first. A long trade is started by buying with the expectation that the price will rise, in order to sell at a higher price in the future and realize a profit. However, as the Binance exchange does not support short trading and all historical data was retrieved from it, implementing a system with short-trading capabilities would be unrealistic. Accordingly, short-trading was not considered in this work's system, only long-trading is executed.

There are several tools to analyse different markets, but the two major categories are Fundamental and Technical analysis. These two approaches to analyse and forecast the market are not exclusive, they may be applied together and attempt to determine the direction prices are likely to move [21].

Fundamental analysis focuses on the key economic forces of supply and demand that may cause prices to increase, decrease or stay level [21]. The essence of fundamental analysis, is that the actual prices of securities tend to move towards their intrinsic values [22]. Accordingly, this approach intends to determine an asset's intrinsic value taking into account economic, financial, industry, company conditions and other relevant aspects that may affect the asset's value, such as the quality of management [23]. In this work, the cryptocurrency exchange market is analysed. This market is in its infancy, meaning there is a lack of track records to analyse. Additionally cryptocurrencies are not corporations, but rather commonly a decentralized representation of value. Thus, a traditional fundamental analysis cannot be performed for this market. Besides, for the cryptocurrency markets, it has been found in literature, mentioned in Sect. 2.3.1, that technical indicators typically offer a best predictive results relatively to fundamental indicators. Besides, this type of data can be misdated, often multiple corrections to the initial values are issued, data is usually less reliable due to self-interest or company policies, etc. [24, 25]. Lastly, this type of analysis is typically used for long-term trading, not short-term trading (such as hourly, daily or weekly) [26]. Accordingly, only technical analysis was employed in this work. With this in mind, the next section is dedicated to introducing technical analysis and the utilized set of technical indicators.

It's worth adding that these two categories of analysis reject both the Random Walk Theory and the Efficient Markets Hypothesis. If markets were random or entirely efficient, no forecasting technique could work.

2.2.1 Technical Analysis

Technical analysis focuses on the analysis of past (or historic) price patterns in order to forecast market behaviour [27]. Unlike fundamental analysis knowing the cause of market movements is not required, only the effect itself. What matters is past trading data and what information might this data provide about future price movements. This type of analysis is based on three premises [21]:

- Market action discounts everything: Anything that may affect the price is already reflected in the price of the market. It is assumed that fundamental factors are utilized in an indirect way as current price reflects all information contained in the past;
- Price moves in trends: Prices evolve in a non-linear fashion over time. The non-linearities, however, contain certain regularities or patterns [28]. The price of a market is more likely to follow a previous trend rather than moving in a totally unpredictable manner. Hence, it's worth identifying the patterns and trends in early stages of their development for the purpose of trading according to their direction.
- History repeats itself: Technical analysis and the study of market action are both deeply related to the study of human psychology, which tends to react the same way in similar market conditions.

Mathematical metrics, named technical indicators, can be developed in order to summarize relevant information of historic data and volume of a financial time series into short-term statistics with the objective of forecasting the value, direction or behaviour of an asset.

An important advantage of using technical analysis is that it can be simplified into a pattern recognition problem in which the inputs are historical prices and technical indicators, while the outputs are the predictions of the future market movements estimated from past data [29]. This work's main concern is making an accurate prediction of future market movements with time sampling frequencies in the order of the minutes, thus, as previously stated, only technical analysis will be used.

The remainder of this section will be dedicated to explaining in further detail the technical indicators applied on the historical data of cryptocurrency markets as well as the motivations for their choice. A more elaborate description of each technical indicator can be found in [21, 23, 30].

- **Moving Averages**: The moving average is one of the most versatile and widely used technical indicator. It is used to dampen the effects of short term oscillations, through a smooth line representing the successive average, usually of prices. In a moving average, a body of data to be averaged moves forward with each new trading period. Old data is dropped as new data becomes available which causes the average to move along the time scale. By definition, this indicator is based on past prices. A moving average is a trend following indicator, it is unable to anticipate, only to react, thus, the moving average is a lagging indicator: it is able to follow a market and announce that a trend has begun, but only after the fact.
The two most widely used types of moving average are Simple Moving Average (SMA) and Exponential Moving Average (EMA), whose equations are represented in Eqs. (2.1) and (2.2) respectively.

$$SMA_p(n) = \sum_{i=p-n}^{p} \frac{Close_i}{n}. \tag{2.1}$$

In Eq. (2.1), n refers to the number of time periods in the body of data to be averaged and p refers to the current period. The SMA is formed by calculating the simple average closing price over a specific period of time.

$$EMA_p(n) = EMA_{p-1} + \left(\frac{2}{n+1}\right)[Close_p - EMA_{p-1}]. \tag{2.2}$$

In Eq. (2.2), once again, n refers to the number of time periods in the body of data to be averaged and p refers to the current period. Note that the first value coincides with the closing value.
Comparing Eq. (2.1) to Eq. (2.2), it can be noted that the major difference between these two types of moving average, is that EMA assigns a greater weight to recent data, giving the ability to react faster to recent price variations.

- **MACD**: The Moving Average Convergence-Divergence is a simple momentum oscillator technique calculated using the difference between two exponential moving averages. To calculate the MACD line, traditionally, a 26-period EMA of the price is subtracted from a 12-period EMA, also of the price. These time periods are adjustable though. The MACD line is usually plotted at the bottom of a price chart along with the signal line. The signal line is an EMA of the MACD, commonly a 9-period EMA is used. Finally, the difference between the two former lines composes the MACD histogram. In Eq. (2.3) the three components of the MACD indicator are presented.

$$MACD = EMA_{12} - EMA_{26}, \tag{2.3a}$$

$$Signal\,Line = EMA_9(MACD),\tag{2.3b}$$

$$MAC\,Dhistogram = MACD - Signal\,Line.\tag{2.3c}$$

The real value of the histogram is spotting whether the difference between the MACD and signal line is widening or narrowing. When the histogram is positive but starts to fall toward the zero line, the uptrend is weakening. Conversely, when the histogram is below negative but starts to move upward towards the zero line, the downtrend is losing its momentum. Although no actual buy or sell signal ought to be given until the histogram crosses its zero line, the histogram turns provide earlier warnings that the current trend is losing momentum. The actual buy and sell signals are given when the MACD line and signal line cross, that is, when the histogram is zero. A crossing by the MACD line above the signal line can be translated into a buy signal. The opposite would be a sell signal. Histogram turns are best used for spotting early exit signals from existing positions.

- **RSI**: The Relative Strength Index is a momentum oscillator that measures the speed and change of price movements. This technical indicator is used to evaluate whether a market is overbought or oversold. The formula used on its calculation is:

$$RSI(n) = 100 - \frac{100}{1 + RS(n)}, \quad \text{with} \quad RS(n) = \frac{AverageGains}{AverageLosses}.\tag{2.4}$$

In Eq. (2.4), n refers to the number of time periods being analysed (traditionally 14 time periods are used), $AverageGains$ refers to the average gain of up periods during the last n periods and $AverageLosses$ refers to the average loss of down periods during the last n periods. The RSI varies between a low of 0 (indicating no up periods) to a high of 100 (indicating exclusively up periods). Traditionally, movements above 70 are considered overbought, while an oversold condition would be a move under 30. A RSI divergence with price is a warning of trend reversal.

Figure 2.1 contains the OHLC chart and the corresponding RSI line for an excerpt of the GOBTC currency pair around October 23rd, 2018 with 1-min precision. The visible RSI red overbought area (above 70) and green oversold line (under 30) represent respectively selling and buying opportunities.

- **ROC**: The Rate Of Change is a simple momentum oscillator used for measuring the percentual amount that prices have changed over a given number of past periods. Traditionally 10 time periods are used. A high ROC value indicates an overbought market, while a low value indicates an oversold market. The formula for calculating this indicator is as follows:

$$ROC(n) = 100 \times \frac{Close_p - Close_{p-n}}{Close_{p-n}}.\tag{2.5}$$

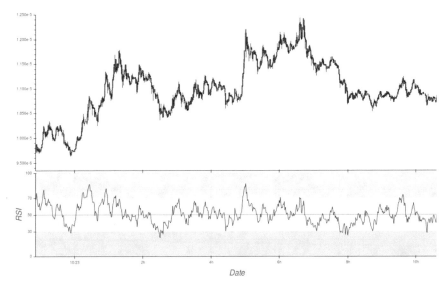

Fig. 2.1 Section of *GOBTC* currency pair OHLC prices with its respective RSI plotted underneath

In Eq. (2.5), *n* refers to the number of time periods being analysed, and *p* corresponds to the current period.

- **Stochastic Oscillator**: The Stochastic Oscillator's intent is to determine where the most recent closing price is in relation to the price range of a given time period. Three lines are used in this indicator: the %K, the fast %D line and the slow %D line. The %K line, the most sensitive of the three, simply measures percentually where the closing price is in relation to the total price range for a selected time period, typically of 14 periods. The second line, the fast %D is a simple moving average of the %K line, usually of 3 periods. The previously mentioned %K line compared with a three-period simple moving average of itself, the fast %D line, corresponds to the fast stochastic. For the fast stochastic, when the %K line is above the %D line, an upward trend is indicated. The opposite indicates a downward trend. If the lines cross, the trend is losing momentum and a reversal is indicated. However, the %K line is too sensitive to price changes and due to the erratic volatility of the fast %D line, many false signals occur with rapidly fluctuating prices. To combat this problem, the slow stochastic was created. The slow stochastic consists on comparing the original fast %D line with a 3-period simple moving average smoothed version of this same line, called slow %D line. In other words, the slow %D line is a doubly smoothed moving average of the %K line.

The formulas for the %K and both %D lines are as follow:

$$\%K_n = 100 \times \frac{Close_t - min(Low)_n}{max(High)_n - min(Low)_n}; \tag{2.6a}$$

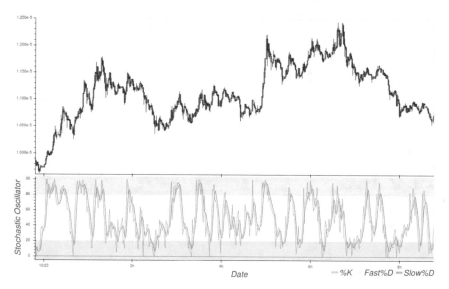

Fig. 2.2 Section of *GOBTC* currency pair OHLC prices with its respective %K, fast %D and slow %D stochastic oscillator lines plotted underneath

$$Fast\%D = SMA_p(\%K_n);\tag{2.6b}$$

$$Slow\%D = SMA_p(Fast\%D).\tag{2.6c}$$

In Eq. (2.6), $Close_t$ corresponds to the current close, $min(Low)_n$ refers to the minimum Low of the previous n periods, and $max(High)_n$ refers to the maximum High of the previous n periods. SMA_p corresponds to a simple moving average of p periods. Traditionally 14 periods are used for calculating the %K line, and 3 periods are used in the simple moving average of the %D line.

In Fig. 2.2 an excerpt of the GOBTC price for October 23rd, 2018 with 1-min precision is represented along with the respective %K, fast %D and slow %D lines of the stochastic oscillator. Low values of the stochastic oscillator (under 20) indicate that the price is near its local low during the respective time period, conversely, high values (over 80) indicate that the price is in a local high during the respective time period.

- **CCI**: The Commodity Channel Index is an oscillator used to measure the variation of a price from its statistical mean. A high CCI value indicates that prices are unusually high compared to the average price, meaning it is overbought. Whereas a low CCI value indicates that prices are unusually low, meaning it is oversold. Traditionally, high and low CCI values respectively correspond to over 100 and under -100. The formula for calculating this indicator is as follows:

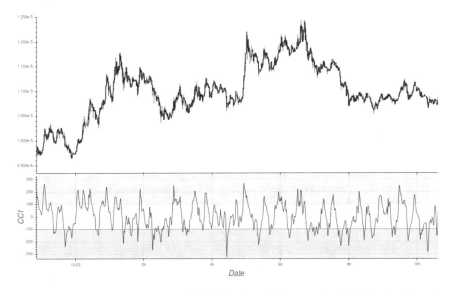

Fig. 2.3 Section of *GOBTC* currency pair OHLC prices with its respective CCI plotted underneath

$$CCI(n) = \frac{1}{0.015} \times \frac{TP_p - SMA_n(TP_p)}{\sigma_n(TP_p)}, \quad \text{with} \ \ TP_p = \frac{High_p + Low_p + Close_p}{3}, \quad (2.7)$$

where TP_p is referred to as the typical price and $High_p$, Low_p and $Close_p$ represent the respective prices for the time period p. The item $SMA_n(TP_p)$ is the simple moving average of the typical price for the previous n periods under consideration, and $\sigma_n(TP_p)$ corresponds to the mean deviation of the SMA during the previous n periods. Commonly, 14 periods are used for the SMA used in this indicator. Lambert, the creator of this indicator, set the constant 0.015 for scaling purposes, to ensure that approximately 70–80% of CCI values would fall between -100 and $+100$.

An excerpt of the GOBTC currency pair for October 23rd, 2018 is charted with a 1-min precision in Fig. 2.3. Low CCI values (under -100) indicate that the current price is under the average price, while conversely, high CCI values (over 100) indicate that the current price is above the average price.

- **OBV**: The On Balance Volume indicator is a running total of volume. It relates volume to price change in order to measure if volume is flowing into or out of a market, assuming that volume changes precede price changes.

  The total volume for each day is assigned a plus or minus sign depending on whether the price closes higher or lower than the previous close. A higher close causes the volume for that day to be given a plus value, while a lower close counts for negative volume. A running cumulative total is then maintained by adding or subtracting each day's volume based on the direction of the market close. The formula used on its calculation is:

$$OBV_p = \begin{cases} OBV_{p-1} + Volume_p, & \text{if } Close_p > Close_{p-1} \\ OBV_{p-1} - Volume_p, & \text{if } Close_p < Close_{p-1} \\ OBV_{p-1}, & \text{if } Close_p = Close_{p-1} \end{cases} \quad (2.8)$$

In Eq. (2.8), p refers to the current period and $p - 1$ refers to the previous period. The OBV line should follow the same direction of the price trend. If prices show a series of higher peaks and troughs (an uptrend), the OBV line should do the same. If prices are trending lower, so should the OBV line. It's when the OBV line fails to move in the same direction as prices that a divergence exists and warns of a possible trend reversal. It is the trend of the OBV line that is relevant, not the actual numbers themselves.

- **ATR**: The Average True Range indicator is used as a measurement of price volatility. Strong movements, in either direction, are often accompanied by large ranges (or large True Ranges). Weak movements, on the other hand, are accompanied by relatively narrow ranges. This way, ATR can be used to validate the enthusiasm behind a move or breakout. A bullish reversal with an increase in ATR would show strong buying pressure and reinforce the reversal. A bearish support break with an increase in ATR would show strong selling pressure and reinforce the breaking of support. To calculate the ATR, the True Range must be firstly calculated:

$$TR_p = max\{High_p - Low_p; \; |High_p - Close_{p-1}|; \; |Low_p - Close_{p-1}|\}. \quad (2.9)$$

In Eq. (2.9), p refers to the current period and $p - 1$ refers to the previous period. Having calculated the True Range, the next step is calculating the Average True Range. The ATR is a simple average of the previous n (traditionally 14 time periods are used) True Range values:

$$ATR_p(n) = \frac{ATR_{p-1} \times (p - 1) + TR_p}{n}. \quad (2.10)$$

In Eq. (2.10), p refers to the current period, $p - 1$ refers to the previous period and n refers to the number of time periods to be analysed.

Figure 2.4 contains the OHLC chart and the corresponding ATR line for an excerpt of the GOBTC currency pair around October 23rd, 2018 with 1-min precision. It can be observed that whenever larger variations in price happen (specially around the 6h mark when a sharp increase is followed by a sharp decrease), the ATR value increases accordingly indicating an increase in volatility, while on sideways periods, the ATR value is low indicating reduced volatility.

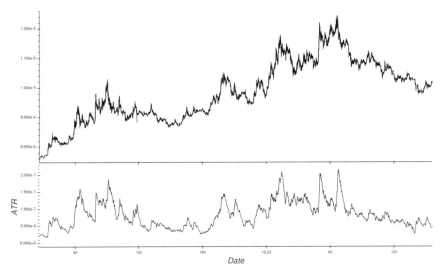

Fig. 2.4 Section of *GOBTC* currency pair OHLC prices with its respective ATR plotted underneath

2.2.2 Time Series Forecasting

A time series is a sequence of observations taken sequentially in time. Its analysis, modelling and forecasting is a subject of study in many studies through different fields of research.

Analysing financial time series is extremely challenging due to the dynamic, non-linear, non-stationary, noisy and chaotic nature of any market [27]. Analysis of financial markets is carried out as an attempt to find the best entry (buy) and exit (sell) points to gain advantage over the market, increasing gains and minimizing the risk.

The key element for better financial decision making is better forecasting [31]. In this work it is assumed that financial time series do not follow a random walk, so forecasting these time series can be translated into a mathematical and probabilistic problem. Forecasting basically is understanding which variables lead to predict other variables [32]. This requires having a clear understanding of the timing of lead-lag relations among many variables, understanding the statistical significance of these relations and learning which variables are the most relevant to be observed as signals for predicting the market moves.

To forecast financial time series, typically machine learning procedures are employed. These procedures are capable of analysing large quantities of seemingly noisy and uncorrelated data in order to detect patterns and predict future data, or to perform other kinds of decision making under uncertainty. Moreover, this approach provides a reaction time notably faster than a human investor could offer.

In this thesis a major objective is using technical indicators as the available data to describe, or forecast, the dichotomous event: will a specific currency pair be bullish or otherwise (bearish or sideways) in the next instant of a time series. To achieve this aim, several supervised learning classification approaches are suggested. The main goal of supervised learning is to create a model out of labelled training data in order to make predictions about unseen or future data. The term supervised refers to a set of samples where the desired output signal is priorly known. Classification is a subcategory of supervised learning where the desired output signals, are discrete, commonly known as class labels.

The applied classification methods are responsible for finding the best fitting and most parsimonious model to describe the relationship between a dependent response variable, or outcome, and a set of independent predictor variables, also called covariates, features or attributes. The independent variables are the characteristics which are potentially related to the outcome variable of interest. After a model has been fit, it may be used to predict new future data using only the set of independent predictor variables.

In this thesis, four total multivariate[3] learning methods were used, two of them, the *Logistic Regression* and the *Support Vector Machine* methods are linear and the other two, the *Random Forest* and the *Decision Tree Gradient Boosting* methods are non-linear. In the end a combined solution, an ensemble of these 4 algorithms, is calculated. This section contains a brief summary of the classification models utilized in this study. However, before this, a few concepts about machine learning, taken into consideration throughout this thesis, are introduced.

A common problem in machine learning that may arise is called overfitting, this is when a model performs well on training data but does not generalize well to unseen test data. Overfitting is often the result of an excessively complex model. This problem is present due to high variance, which is the tendency of being overly sensitive to small fluctuations in the training set. High variance can cause an algorithm to model the random noise in the training data, rather than the intended outputs, causing overfitting.

On the other side of the spectrum, a model can also suffer from underfitting. Underfitting occurs when a model cannot capture the underlying structure or trend of the data. Underfitting is often the result of an excessively simple model. This occurs when a model has a high bias, which is the tendency to learn erroneous assumptions and relations in the training set. High bias in an algorithm may be responsible for missing relevant relations and patterns between features and target outputs, the main cause of underfitting.

In practice, there is not an analytical way to discover the point where a model is underfit or overfit, instead a balance between these two must be searched by surveying different levels of complexity. Different strategies used by each learning algorithm to achieve an ideal bias-variance trade-off will be presented in the following subsections.

---

[3]More than two features are analysed together for any possible association or interactions.

2.2.2.1 Logistic Regression

A binomial logistic regression (LR) [33] is used to model a binary dependent variable. Logistic Regression distinguishes itself from linear regression because the outcome variable is binary or dichotomous. Apart from this, the methods utilized in linear and logistic regression are similar.

In this type of learning algorithm, a single outcome variable Y_i follows a Bernoulli probability function that takes the outcome of interest, usually represented with the class label $Y = 1$, with probability p while the unwanted outcome, usually represented with the class label $Y = 0$, has probability $1 - p$.

The odds in favour of a particular outcome can be represented as $\frac{p}{(1-p)}$, where p stands for the probability of the wanted outcome. The logit function is simply the logarithm of the odds ratio, also known as the log-odds:

$$logit(p) = log\left(\frac{p}{1 - p}\right). \tag{2.11}$$

This logit function is capable of transforming input values in the range of 0–1, to values over the entire real number range, which can be used to express a linear relationship between feature values and the log-odds as such [34]:

$$logit(P(Y = 1|x)) = \beta_0 + \sum \beta_i x_i, \tag{2.12}$$

where $P(Y = 1|x)$ is the conditional probability of Y belonging to class label '1' given x_i, the feature values, β_0 is the intercept (point that intercepts the function in the Y axis) and β_i corresponds to the coefficient associated with each respective feature.

Joining Eqs. (2.11) and (2.12), the conditional probability can be represented as:

$$P(Y = 1|x) = \frac{1}{1 + exp(-\beta_0 - \sum \beta_j x_{ij})}. \tag{2.13}$$

Equation (2.13) is called the logistic or sigmoid function (due to its S-shape). From this function it can be seen that $P(Y = 1|x)$ varies between 0 (as x approaches $-\infty$) and 1 (as x approaches $+\infty$). Thus, it is clear that the logistic function is able of transforming any real input into the range of 0–1. The class probabilities are obtained as such. The last aspect to cover is how the coefficient values are defined.

LR models are usually fit by maximum likelihood [33]. An iterative process that yields values for the unknown parameters that maximize the probability of obtaining the observed set of data. The principle of maximum likelihood states that the best estimate for the set of coefficients, is the set that maximizes the *log-likelihood*. This is equivalent to minimizing the negative log-likelihood [33]. In this work the optimization problem utilized to obtain the coefficients and intercept, minimizes the following cost function:

$$\min_{\beta, \beta_0} \frac{\beta^2}{2} + C \sum_{i=1}^{n} log(exp(-Y_i(x_i^T \beta + \beta_0)) + 1), \qquad (2.14)$$

where $\frac{\beta^2}{2}$ is the L2 regularization penalty, C is a parameter inverse to the regularization strength and $\sum_{i=1}^{n} log(exp(-Y_i(x_i^T \beta + \beta_0)) + 1)$ corresponds to the negative
log-likelihood equation [35]. In the negative log-likelihood equation, β represents
the coefficients associated with each respective feature value, x_i, both are in a vector
structure.

A method of finding a good bias-variance trade-off for a model is by tuning its
complexity via the regularization strength parameter, C. This method can handle
collinearity (high correlation among features), filter out noise from data and prevent
overfitting [34]. The concept behind regularization is introducing additional information (bias) to penalize extreme parameter weights. Hence, smaller C values specify
stronger regularization, that is, the coefficients jointly shrink. The ideal value of C is
supposed to generate a model that generalizes well to new, previously unseen data.

2.2.2.2   Random Forest

A Random Forest (RF) [36] is a method of ensemble learning where multiple classifiers are generated and their results are aggregated. Random Forest is a modification
of the method bootstrap aggregating (bagging) [37] of classification trees.

Before all else, a classification or decision tree is a simple model that takes into
account the whole dataset and all available features. Briefly, a decision tree utilizes
a top-down approach where binary splits are continuously repeated from the root
node until a certain stopping criteria is met, such as maximum depth (path length
from a root to a leaf) or until the minimum number of samples required to split an
internal node is no longer surpassed. Its construction starts with the total training
dataset instances, all possible ways of creating a binary split among those instances
based on input features are evaluated.

Each split chooses the best feature, and the best value for that feature to produce
the most meaningful separation of the target label. Different metrics can be used to
evaluate the quality of a split. The most common approach to splitting nodes is to
define a measure of the impurity of the distribution at a node, and choose the split
that reduces the average impurity the most [38]. Among several, a commonly used
splitting criteria in binary decision trees is the Gini index [39]. Gini Index is a metric
to measure how often a randomly chosen element would be misclassified.

Having found the 'best' split according to a metric, a node in the tree partitions
training instances either left or right according to the value of the feature. The subsets
of training instances may be recursively split to continue growing the tree until a
maximum depth, or until the quality of the splits is under a certain threshold. Each leaf
of the decision tree contains predictions for a target class label. Decision trees possess

many hyper-parameters[4] [40] that may be tuned and are taken into consideration in a random forest model as well. These trees tend to have high variance and overfit on training data, leading to a poor generalization ability on unseen data. Methods such as pruning can be used to avoid overfitting on the final tree [35].

In the bagging method however, multiple decision trees are independently constructed using random samples drawn with replacement (known as a bootstrap sample) of the data set in order to reduce the variance. In the end, a simple majority vote among all constructed trees is undertaken for prediction [39]. Bagging is capable of reducing overfitting while increasing the accuracy of unstable models [37]. However, re-running the same learning algorithm on different subsets of the data may still result in correlated predictors, hence, capping the reduction of variance.

Random Forests improves the variance reduction on bagging by reducing the correlation between trees [39]. In order to do so, an additional layer of randomness is added to the bagging procedure. Instead of using all the available features, only a random subset of features is used to grow each tree of the forest. Contrarily to standard classification trees where each node is split using the best split among all variables, in a random forest, each node is split using the best among a subset of predictors randomly chosen at that node. This strategy turns out to be robust against overfitting [41]. Reducing the amount of features will reduce the correlation between any pair of trees in the ensemble, hence, the variance of the model is reduced [36].

Let N be the number of data points in the original dataset, briefly, each tree in the random forest algorithm is created as follows [39]:

1. Draw a random sample of size N with replacement (hence) from the original data (bootstrap sample);
2. Grow a random forest tree to the bootstrapped data. Until the minimum node size is reached, recursively repeat the following steps for each terminal node of the tree:

   a. Select a fixed amount of variables at random from the whole set;
   b. Split the node using the feature that provides the best split according to the objective function;
   c. Split the node into two daughter nodes.

Note that the size of the bootstrap sample is typically chosen to be equal to the number of samples in the original dataset as it provides a good bias-variance trade-off [34]. Each time the previous steps are repeated a new tree is added to the ensemble. Repeating this processes multiple times outputs an ensemble of trees with as many trees as this process was repeated. After a Random Forest is generated, a classification can be obtained from doing a majority voting of the class vote from each individual tree in the forest. The predicted class probabilities of an input sample are computed as the mean predicted class probabilities of all trees in the forest. The probability estimate of a tree for a given input sample is the class probability of the corresponding

---

[4]Parameter whose value is set before the learning process begins as it is not directly learnt within the estimator.

terminal node. In turn, the class probabilities for a terminal node are estimated by the relative frequency of the class of interest in that terminal node.

The two main parameters in random forest are the number of trees in the ensemble and the number of variables to be considered when looking for the best split [34]. Typically, with a larger number of trees, the performance of the random forest classifier is improved, but at the expense of an increased computational cost. The number of variables taken into consideration on each split is commonly the square root of the amount of variables.

2.2.2.3 Gradient Decision Tree Boosting

Gradient Decision Tree Boosting (GTB) in this work is utilized through the XGBoost framework, a scalable machine learning system for tree boosting [42].

Boosting is a general method for improving the accuracy of any given learning algorithm [43]. Boosting is the process of combining many *weak classifiers* with limited predictive ability into a single more robust classifier capable of producing better predictions of a target [44]. A weak classifier is one whose error rate performs only slightly better than random guessing, hence these learners by definition have a high bias. Boosting is an ensemble method very resistant to overfitting that creates each individual members sequentially [35]. The newest members are created to compensate for the instances incorrectly labelled by the previous learners. To accomplish this, misclassified data increases its weight to emphasize the most difficult cases. Simply put, the main goal of boosting is reducing the error of the weak learning algorithms [45].

Gradient boosting replaces the potentially difficult function or optimization problem existent in boosting. It represents the learning problem as a gradient descent on some arbitrary differentiable loss function in order to measure the performance of the model on the training set [46].

The previously mentioned decision tree algorithm has many advantages (such as being easily interpretable, immune to outliers, etc.), nonetheless, they seldom provide an adequate level of predictive accuracy [39]. Decision tree gradient boosting means to improve the accuracy of decision trees, the weak classifiers, while maintaining most of their desirable properties for data mining.

In this thesis a generalised gradient boosting implementation that includes a regularization term (used to combat overfitting) and support for arbitrary differentiable loss functions is used. Instead of optimising a plain squared error loss, an objective function consisting of a loss function over the training set and a regularization term used to penalize the complexity of the model is applied as follows:

$$Obj = \sum_i L(y_i, \hat{y}_i) + \sum_k \Omega(f_k). \tag{2.15}$$

In Eq. (2.15), the first term, $L(y_i, \hat{y}_i)$, can be any convex differentiable loss function that measures the difference between the predicted label $\hat{y}_i$ and its respective true label y_i for a given instance. In this thesis's proposed system the log-likelihood loss will be used as loss function. Through the usage of this loss function, the calculation of probability estimates is enabled. Combining the principles of decision trees and logistic regression [47], the conditional probability of Y given x is obtained as follows [48]:

$$P[Y = 1|x] = \frac{e^{F(x)}}{e^{F(x)} + e^{-F(x)}}, \tag{2.16}$$

where $F(x)$ corresponds to the sum of the weights of the terminal leafs for a given sample, i.e., $F(x)$ is the weighted average of base classifiers taking into account in the whole space of trees.

The second term of Eq. (2.15), $\Omega(f_k)$, is used to measure the complexity of a tree f_k and is defined as:

$$\Omega(f_k) = \gamma T + \frac{\lambda ||w||^2}{2}, \tag{2.17}$$

where T is the number of leaves of tree f_k and w is the leaf weights (i.e. predicted values stored at the leaf nodes). Including Eq. (2.17) in the objective function (Eq. (2.15)) forces the optimization of a less complex tree, which assists in reducing overfitting. The second term of this equation, $\frac{\lambda ||w||^2}{2}$, corresponds to the L2 regularization utilized previously in LR. λ is the L2 regularization strength and γT provides a constant penalty for each additional tree leaf, both are configurable [49].

Boosting proceeds in an iterative manner, hence the objective function for the current iteration m in terms of the prediction of the previous iteration $\hat{y}_i^{(m-1)}$ is adjusted by the newest tree f_k:

$$Obj^m = \sum_i L(y_i, \hat{y}_i^{(m-1)} + f_k(x_i)) + \sum_k \Omega(f_k) \tag{2.18}$$

Equation (2.18) can be optimised to find the f_k which minimises the objective. Simplifying the previous equation outputs a measure of quality or score of a given tree [49]. This score is similar to the impurity score for evaluating decision trees, except that it is derived for a wider range of objective functions.

Lastly, in order to further prevent overfitting the techniques shrinkage and feature sub-sampling were added. Shrinkage [46] is a technique to scale newly added weights by a factor after each step of tree boosting. This technique reduces the individual influence of each tree allowing future trees to improve the model. The feature sub-sampling concept has already been mentioned in the random forest section. This technique specifies the fraction of features to be randomly considered when constructing each tree [36].

Fig. 2.5 Hyperplane
dividing two linearly
separable classes

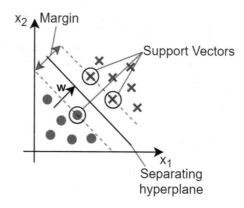

2.2.2.4 Support Vector Machine

A Support Vector Classifier (SVC) [50] is a classifier algorithm whose primary objective is maximizing the margin, of a separating hyperplane (decision boundary) in an n-dimensional space, where 'n' coincides with the number of features used. The margin corresponds to the distance between the separating hyperplane and the training samples that are closest to this hyperplane, the support vectors. Figure 2.5 provides an illustration of these concepts for a simple case with only two features (x_1 and x_2), two classes (one class is represented with blue crosses and the other class with red circles). Vector **w** is the normal to the hyperplane.

The hyperplane is supposed to separate the different classes, that is, in a binary classification problem, the samples of the first class should stay on one side of the surface and the samples of the second class should stay on the other side. In Fig. 2.5 the data is linearly separable, i.e., a separating linear hyperplane is able to perfectly separate the two classes. However, unlike the case illustrated in Fig. 2.5, in many occasions, the utilized data may not be fully linearly separable. In this work, it is most likely that the data is not entirely linearly separable.

Decision boundaries with large margins tend to have a lower generalization error of the classifier, whereas models with small margins are more prone to overfitting, hence, it is important to maximize the margins [34]. With the purpose of achieving a better generalization ability, margins may be tampered, but at the cost of possibly increasing the odds of misclassifying outlier data points. For this purpose, a slack variable, ξ, indicating the proportional amount by which a prediction is misclassified on the wrong size of its margin is introduced. This formulation, called soft-margin SVM [50], enables controlling the width of the margin and consequently can be used to tune the bias-variance trade-off. The positive slack variable is introduced in the hyperplane equations to allow for misclassified points as such [51]:

$$\mathbf{w} \cdot \mathbf{x}_i + b \geq 1 - \xi_i \text{ where } \xi_i \geq 0 \, \forall_i \text{ for } y_i = 1; \tag{2.19a}$$

$$\mathbf{w} \cdot \mathbf{x}_i + b \leq \xi_i - 1 \text{ where } \xi_i \geq 0 \, \forall_i \text{ for } y_i = -1, \tag{2.19b}$$

where $\mathbf{w}$ is the normal to the hyperplane, $\mathbf{x}_i$ contains the input features, b is the offset of the hyperplane commonly called bias and ξ, as previously mentioned is the slack variable. Equations (2.19) can be combined into:

$$Y_i(\mathbf{w} \cdot \mathbf{x}_i + b) \geq 1 - \xi_i \text{ where } \xi_i \geq 0 \forall_i, \tag{2.20}$$

where Y_i is the binary output class label. With this soft-margin formulation, data points on the incorrect side of the decision boundary have a penalty that increases with the distance from the margin.

In order to maximize the margin, the hyperplane has to be oriented as far from the support vectors as possible. Through simple vector geometry this margin is equal to $\frac{1}{||\mathbf{w}||}$ [39], hence maximizing this margin is equivalent to finding the minimum $||\mathbf{w}||$. In turn, minimizing $||\mathbf{w}||$ is equivalent to minimizing $\frac{1}{2}||\mathbf{w}||^2$, the L2 regularization penalty term previously utilized in LR. The use of this term enables quadratic programming which is computationally more convenient [39]. Therefore, in order to reduce the amount of misclassification, the objective function can be written as follows:

$$\min_{\mathbf{w},b,\xi} \frac{||\mathbf{w}||^2}{2} + C \sum_{i=1}^{n} \xi_i \text{ subject to } Y_i(\mathbf{w} \cdot \mathbf{x}_i + b) \geq 1 - \xi_i, \xi_i \geq 0, i = 1, \ldots, n$$
$$\tag{2.21}$$

where $||\mathbf{w}||^2$ is the squared norm of the normal vector $\mathbf{w}$. C is the regularization parameter, responsible for controlling the trade-off between the slack variable penalty, ξ, and the size of the margin, consequently tuning the bias-variance trade-off. A larger regularization parameter focuses attention on (correctly classified) points near the decision boundary, while a smaller value focuses on data further away [39].

In order to calculate probability estimates, Platt scaling [52] is employed [53]. Platt scaling first requires regularly training the SVM and then optimizing parameter vector A and intercept point B such that:

$$P(Y = 1|x) = \frac{1}{1 + exp(Af(x) + B)}, \tag{2.22}$$

where $f(x)$ is the signed distance of a sample from the hyperplane and $P(Y = 1|x)$, similarly to Eq. (2.13) is the conditional probability of a given output Y belonging to class label '1' given x, the feature values.

Effectively, Platt scaling trains a probability model on top of the SVM's outputs under a cross-entropy loss function. To prevent this model from overfitting, an internal five-fold cross validation is used. This procedure is clearly expensive computationally but it is necessary in order to obtain the probability estimates used to later create the unweighted average.

In this work a linear SVM is utilized, but it is possible that a non-linear classifier could provide a better accuracy. Kernel methods can deal with such linearly inseparable data by creating non-linear combinations of the original features to project them onto a higher dimensional space via a mapping function.

Kernels methods operate in a high-dimensional, implicit feature space without computing the coordinates of the data for that space, but rather by computing the inner products between the images of all pairs of data in the feature space [34]. This kernel approach is a computationally efficient approach for accommodating a non-linear boundary between classes [54]. Additional kernel methods, such as Polynomial or Radial Basis Function can be further investigated in [54].

2.2.2.5   Ensemble Voting

The goal behind ensemble voting (EV) is to combine different classifiers into a meta-classifier with better generalization performance than each individual classifier alone. The weaknesses of one method can be balanced by the strengths of others by achieving a systematic effect. In fact, a necessary condition for an ensemble of classifiers to become more accurate than any of its individual members is if the classifiers are accurate and diverse [55, 56].

If each classifier's error rate is better than random, increasing diversity in classifications between classifiers tends to decrease the test error as the error concentration is diluted to less than the majority of votes for any given test case to be classified. Hence, ensemble voting tends to result in the correct classification [57].

In this thesis, a heterogeneous ensemble[5] was combined in a linear manner: the prediction's probability estimates from each different individual classifier were combined according to a simple unweighted average, giving equal probability to each individual output. This process, also known as soft majority voting [56], is the reason why all previous learning algorithms ought to yield a probability estimate for each class label, often with the cost of additional computer complexity.

2.3   State-of-the-Art

Having described the essential background for this thesis, a study of previous papers dedicated to forecasting financial markets using technical analysis are overviewed in this section. Special attention will be given to papers using the same learning algorithms used in this thesis (LR, RF, GTB, SVM and Ensemble) to serve as a baseline for this thesis to be compared.

---

[5]Ensemble containing different learning techniques.

2.3.1 Works on Machine Learning and Financial Markets

Special attention will be given to the cryptocurrency market, however, as this market is relatively new, only Bitcoin, and somewhat Ethereum, which are the most popular currencies, have been in some measure examined in regards to market forecasting (as well as to the many remaining details of these cryptocurrencies, such as security, its architecture, criminality and ethics, etc.), the remaining cryptocurrencies are relatively disregarded in literature. However, the underlying principle in forecasting financial time series using technical analysis is the same for all sort of financial markets.

Firstly, a couple proposals aimed at improving forecasting financial time series employed in this work are introduced. De Prado in [24], rather than using, as is customary, fixed time interval bars (minute, hour, day, week, etc.), proposes the usage of a different type of price bar. Fixed time interval bars often exhibit oversampling, which leads to penalizing low-activity periods and undersampling penalizing high-activity periods, as well as poor statistical properties, like serial correlation, heteroscedasticity and non-normality of returns. Instead of time interval bars, as an alternative, the author suggests forming bars as a subordinated process of trading activity. *Volume bars*, for instance, are sampled every time a pre-defined amount of units have been exchanged in the market and *dollar bars* are formed by sampling an observation every time a pre-defined market value is exchanged. According to the author, alternative bars, relatively to the commonly used time bars, achieve better statistical properties, are more intuitive particularly when the analysis involves significant price fluctuations and the research and conclusions tend to be more robust. The author mentions, the concept of resampled interval bars is not common yet in literature. As a matter of fact, throughout literature, no actual experimentation was found that could validate De Prado's claims, Nonetheless, this concept will be experimented in this thesis.

In time series analysis the goal is predicting a series that typically is not deterministic but contains a random component. Statistical forecasting methods are based on the assumption that this random component is stationary in order to make inferences about future observations [58, 59]. Stationary models assume that the process remains in statistical equilibrium with probabilistic properties that do not change over time, in particular it is assumed that a stationary series varies about a fixed constant mean level and has a constant variance [60]. In finance, price series are commonly non-stationary when expressed in their original form. In a time series, the process of counteracting the non-stationarity by considering its change series is called differencing in the time series literature. More formally, $d_t = y_t - y_{t-1}$ is referred to as the first differenced series of y [60]. This type of differencing will be used in this thesis.

To provide a broader perspective of the state-of-the-art, a few examples of financial time series forecasting using alternative learning algorithms than the ones used in this thesis follow:

Cardoso and Neves [61] proposed a system based on genetic algorithms to create an investment strategy intended to be applied on the Credit Default Swaps market.

This market, similarly to the cryptocurrency market is still growing, is subject to speculation and is quite volatile. The employed strategy utilized several instances of genetic algorithms with the objective of increasing profitability. The obtained results suggest that it is possible to create a profitable strategy using only technical analysis as input data, reaching commonly over 50% on return on investments in the CDS market, ergo, it should be possible of acquiring, at the least, matching results for the similar cryptocurrency markets.

Abreu, Neves and Horta [62] proposed using a genetic algorithm to optimize the Naive-Bayes cross validated model output estimation. Using only technical indicators from the Foreign Exchange Market, the proposed architecture improved accuracy of the unoptimized system from 51.39 to 53.95% on the test set and improved the return on investment from 0.43 to 10.29% on the validation set. In addition, an attempt of model visualization using the algorithm t-Distributed Stochastic Neighbour Embedding is done. Shah and Zhang [63] discuss a method of Bayesian regression and its efficacy for predicting price variations only in Bitcoin. The used data consisted of 10-s historical prices and limit order book features. Despite having a decreasing market at the end of their testing period, their strategy was still profitable. The authors claim reaching a 50-day return on investment of 89% and a Sharpe ratio of 4.10.

Jiang and Liang [64] present a model-less convolutional neural network trained using a deterministic deep reinforcement method in order to manage a portfolio of cryptocurrency pairs. Historic data sampled every 30 min from the 12 most-volumed cryptocurrencies of a given exchange is the only input of their system. In this thesis, the selection of cryptocurrencies with the most volume is also utilized to satisfy a set of constraints mentioned by the authors, utilized in this work. The authors obtained an increase of 16.3% in their portfolio value, a sharpe ratio of 0.036 and a maximum drawdown of 29.6%, while the B&H method, used as benchmark, ended with a final portfolio value of 0.87, −1.54 sharpe ratio and 38.2% maximum drawdown. Nadkarni and Neves [65] developed an approach combining Principal Component Analysis with NeuroEvolution of Augmenting Topologies (NEAT) to generate a trading signal. The daily prices, volume and technical indicators from seven financial markets from different sectors were the only inputs. The return on investment was always larger than what was obtained with a B&H strategy. For instance, the *S&P 500* index acquired a rate of return of 18.89%, while B&H achieved 15.71% and the *Brent Crude futures* acquired a rate of return of 37.91% while B&H achieved −9.94%.

Nakano et al. [66] utilize a seven layered artificial neural network (ANN) to create trading signals in the Bitcoin exchange market. Only technical indicators derived from Bitcoin/USD 15-min return data ($return_t = close_t/close_{t-1} - 1$) were used as input. Their system was compared to a benchmark B&H strategy. The authors defined three strategies of generating a trading signal. Two of them entered long and short and long positions but one only used long positions (similarly to this work's strategy). The strategy using only long positions without considering the bid-ask spread, yielded a final ROI of 12.14% while the simple B&H strategy obtained only 2.28%. The two remaining strategies obtained better results, both of them exceeded a 50% ROI. Finally, the authors noted that increasing the amount of technical indicators from 2 (RSI and EMA) to 5 (RSI, EMA, MACD, Stochastics and OBV) originated better results. As a side remark, the authors mention, that all 3 strategies perform well in the period from December 2017 to January 2018, where BTC suffers a substantial drawdown which it's worth praising. Nevertheless, as expected, the strategy that only implemented long positions performed substantially worse relatively to the other two strategies that implement short positions as well.

Tan and Yao [67] reported that a neural network model is applicable to the forecasting of foreign exchange rates. Additionally, the autoregressive integrated moving average (ARIMA) model and a simple B&H were used as benchmarks to compare their results. Two distinct inputs were fed, one contained a time series with weekly data and the other contained technical indicators derived from the time series. Both inputs had 5 different major currency pairs sampled on a weekly basis. It was concluded that technical indicator series yielded the best results in terms of returns. According to the Normalized Mean Squared Error (NMSE) metric used to evaluate the performance of the regression, it may be concluded that Neural Networks yielded better results than both benchmarks.

McNally et al. [68] try to predict the price direction of Bitcoin. For this, a Bayesian optimised recurrent neural network (RNN), a Long Short Term Memory (LSTM) network and ARIMA were explored and compared. The daily historical prices combined with two daily blockchain related features were the only inputs. The LSTM achieved the highest classification accuracy of 52.7% followed by RNN with 50.2% and ARIMA with 50.05%. However the Root Mean Square Error[6] (RMSE) was 6.87% for LSTM, 5.45% for RNN and 53.74% for ARIMA indicating that LSTM and RNN are far superior in forecasting Bitcoin price. The author concluded that Deep Learning models require a significant amounts of data and 1-min precision data would have been used if available while developing the work.

Atsalakis and Valavanis [69] surveyed over 100 published articles focused on neural and neuro-fuzzy techniques derived and applied to forecast stock markets. The authors found out that the majority of all articles implementing data preprocessing found it useful and necessary. Cavalcante et al. [26] more recently, performed a similar comprehensive review of the literature of financial market forecasting, presenting multiple studies applied in various financial markets using a broader set of machine learning techniques.

While this thesis and all the previous mentioned works use historical prices and/or technical indicators, alternative indicators can also be used as input to forecast financial markets, a few examples follow:

Greaves and Au [70] tried to predict whether Bitcoin's price increased or decreased in the next hour using accuracy as the classification metric. In this paper Logistic Regression, Support Vector Machine, Neural Network and a Baseline (percentage of the average price increase) were used. However, the distinctive aspect of this paper are the used inputs. Apart from using "current Bitcoin price", the remaining features were all related to blockchain network features such as "mined bitcoin in

---

[6]The sample standard deviation of the differences between predicted values and observed values (called residuals).

the last hour" or "number of transactions made by new addresses in a given hour". They obtained accuracies of 53.4% for baseline, SVM followed with 53.7%, LR with 54.3% and finally Neural Networks were the best with 55.1%. The authors concluded that using input data from the blockchain alone offers limited predictability, which makes sense as price is dictated by exchanges and these fall outside the range of blockchains. Finally, it is presumed that price related features obtained from cryptocurrency exchanges are the most informative in regards to future price prediction.

Colianni et al. [71] use sentiment analysis to prove whether Twitter data associated with cryptocurrencies may be advantageous to develop a trading strategy. The authors utilized several machine learning methods to predict whether the price of Bitcoin will increase or decrease over a fixed time frame. Over 1 million tweets were analysed by the following machine learning algorithms: Naive-Bayes, LR and SVM. The scores returned from an API for each word's positivity, negativity and neutrality were fed as feature vectors (matrix x) to each learning algorithm in order to obtain a prediction of the next period's Bitcoin price direction. The authors typically obtained better accuracies for daily predictions rather than hourly predictions, but, in both cases accuracies of well over 50% were achieved. According to these results, it seems that a system solely based on twitter data to forecast cryptocurrencies on a minute basis is unfeasible, as great amounts of data would be required. Stenqvist and Lönnö [72] similarly used sentiment analysis to predict whether Bitcoin's value will increase or decrease during a given time period through a time shifted signal obtained from 2.27 million Bitcoin related tweets, processed by a rule-based sentiment analytic software (VADER) able to detect the sentiment intensity. The authors obtained really positive accuracies, however, they concluded that the number of predictions they created was not a significant number, hence venturing into conclusions would be unfounded.

2.3.2 Works on LR, RF, GTB, SVM and Ensemble Methods

A more thorough literature analysis on the five methods employed in this thesis is done in this section.

Ballings et al. [73], start with an extensive literature review, attempt to benchmark the performance of ensemble methods (RF, Adaboost and Kernel Factory) against single classifier models (Neural Networks, LR, SVM and K-Nearest Neighbours) in predicting stock price direction. Historical price data combined with fundamental indicators of thousands of publicly listed European companies was the only input used to forecast one year ahead whether the stock price will increase by a predetermined amount or not. The evaluation metric used is the area under the receiver operating characteristic curve (AUC). The results indicate that RF is the top algorithm followed by SVM, Kernel Factory, AdaBoost, Neural Networks, K-Nearest Neighbours and LR. Ou and Wang [74], likewise, compare the performance of ten forecasting methodologies for the Hang Seng index using only daily historical price data. The best performing methodologies measured in hit ratio[7] are, in descending order, the Least squares support vector machine (LS-SVM) followed by SVM, LR, ANN alongside Gaussian Process, Quadratic discriminant analysis (QDA), Linear discriminant analysis (LDA), Naive-Bayes, decision tree and lastly K-nearest neighbour method. The obtained accuracies are suspiciously high and no explanation is given by the authors, nonetheless, the comparison between these methods seems valid.

Kumar and Thenmozhi [75] compare the performance of Linear Discriminant Analysis (LDA), LR, SVM, RF and ANN in forecasting the direction of change in the next daily closing price for the NIFTY index. The authors suggest that forecasting the direction of the price, rather than price levels, is more effective and generates higher profits. The results were measured by hit ratio, obtaining 56.4%, 59.6%, 62.9%, 67.4% and 68.4% for LDA, LR, ANN, RF and SVM respectively. The superior performance of the SVM is attributed to the fact that it implements a structural risk minimization principle which minimizes an upper bound of the generalization error instead of minimizing the training error. It is also suggested that combining the used models may generate better performances.

Madan et al. [76] used two sets containing Bitcoin's historical price to predict the signal of future price change. The first set had a 10-s sampling precision while the second set had a 10-min precision. The used algorithms were random forests and binomial logistic regression classifiers. The results obtained with the 10-s data were much inferior to the results obtained with the 10-min time intervals. Using 10-min data, resulted in approximately 54% accuracy obtained for LR and 57% for RF on the prediction of the sign of price change in the next (10-min) interval.

Żbikowski in [77] used a set of 10 technical indicators calculated from Bitcoin's historical price (with a 15-min precision) as input to investigate the application of SVM with Box Theory and Volume Weighted in forecasting price direction in the Bitcoin Market with the purpose of creating trading strategies. A simple B&H strategy used as baseline, which obtained a ROI of 4.86%, was outperformed by the BOX-SVM, with 10.6% ROI, and the VW-SVM, with 33.5% ROI.

Mallqui and Fernandes [78] similarly attempt to predict the price direction of Bitcoin, but on a daily basis. The authors, besides the OHLC values and volume, experimented adding several blockchain indicators (total transaction fees, number and cost of transactions and average hash rate), as well as a few "external" indicators (such as crude oil and gold future prices, S&P500 future, etc.). Several attribute selection techniques were employed and always considered the OHLC values and volume as the most relevant attributes. Even though no technical indicators were applied in this work, their usage is suggested as a means for possible improvement. Several ensemble and individual learning methods were experimented in this work. Nonetheless, the best performers were the SVM by itself and an ensemble of a Recurrent Neural Network and a Tree Classifier. The SVM obtained a final accuracy of 59.4% (utilizing 80% of the original dataset dedicated to training) and the ensemble an accuracy of 62.9% (utilizing 75% of the original dataset dedicated to training).

---

[7]Ratio between predicted true positives and the number of real positive cases in the data.

The authors conclude that, with respect to the existent state-of-the art works, their system manages to obtain notably higher accuracies. The results showed that the selected attributes and the best machine learning model achieved an improvement of more than 10%, in accuracy, for the price direction predictions, with respect to the state-of-the-art papers, using the same period of information.

Huang et al. [79] investigate the prediction ability of SVM by forecasting weekly movement direction of NIKKEI 225 index. SVM's performance is compared to LDA, Quadratic discriminant analysis (QDA) and to Elman Backpropagation Neural Networks (EBNN). In the end, the previously mentioned methodologies are combined to form an ensemble classifier. The combined model reached the highest hit ratio of 75% followed by SVM with 73%, EBNN got third place with only 69% hit ratio. It's worth mentioning that only 36 observations were used as test sample which is not statistically significant. Kim [80] compared the performance of SVM to ANN in predicting the daily variation direction of the KOSPI index. The SVM obtained the best hit ratio of 57.8% compared to ANN's hit rate of 54.7%. They concluded that SVM is sensitive to its hyper-parameters, hence it is important to find the optimal values of the parameters.

Akyildirim et al. [81] predict in the most liquid twelve cryptocurrencies utilizing data with different sampling frequencies ranging from daily to minute level. The authors utilized the methodologies SVM, LR, RF and ANN and historical price and technical indicators as inputs. The objective is predicting in a binary form the price direction in the next time step. ANN performed the worse with an accuracy slightly under 55%. It is concluded that no significant gain was acquired from using ANN in the prediction of cryptocurrency returns, however, a larger sample size should be experimented with. LR was consistent across most different time-scales obtaining accuracies averaging 55%. SVM also had low variation across the different time-scales and performed better than LR, obtaining on average accuracies slightly over 56%. Finally, RF obtained the best accuracies at around 59%.

Basak et al. [82] predicted the direction of ten different stock markets prices using RF and GTB using technical indicators obtained from daily historical prices. The predictions are made for time windows between 3 and 90 days. Both methodologies obtained, for 90 days, similar accuracies of over 90% while for 3 days, around 65% was obtained. The suspiciously high accuracies are acknowledged and discussed in this paper. In any case, both algorithms were considered to be robust in predicting the direction of price movements.

Alessandretti et al. [83] test and compare three methods for short-term cryptocurrency forecasting. The first method considers only one single GTB model to describe the change in price of all cryptocurrencies. The second method is built using different GTB models for each cryptocurrency. The third (and last) model is based on the LSTM algorithm. The daily historical price of 1681 cryptocurrencies, as well as a few fundamental indicators, such as market capitalization, market share, age and rank, are used as inputs for each of the three methods. As baseline method, the simple moving average strategy[8] was implemented. The three methods performed better than the baseline, even with fees included. The cumulative returns were positive and plotted for the three methods and the baseline. Between the first two methods, the second was considered to perform better. Nevertheless, both methods performed well for short-term windows, suggesting they exploit short-term dependencies. In the end, the three methods performed better when predictions were based on prices on Bitcoin rather than prices in USD, suggesting that forecasting simultaneously the overall cryptocurrency market trend and the developments of individual currencies is more challenging than forecasting the latter alone. One final remark from this work is that utilizing *price* as an input feature predominantly is attributed a larger relative importance by the predictive models employed, rather than the remaining fundamental indicators.

do Ó Barbosa and Neves [84] in order to predict an economic recession used LR, RF, GBT and an ensemble of the previous three to predict in advance whether the USA economy is in a state of recession or not, hence, the previous classifiers were binary. Technical indicators were the only used input. It was concluded that, individually, RF acquired the best results and GBT was robust no matter the different conditions. However, the classifier obtaining the best Area Under the Curve (AUC) results was the equally weighted average of the 3 classifiers.

Tsai et al. [85] investigate the prediction performance utilizing classifier ensembles to analyse stock returns. Three types of single classifiers were compared to a B&H strategy in this paper: multi-layer perception (MLP) neural network, classification and regression tree (CART) decision trees and LR. Multiple learning methods were used in order to predict whether a given stock yields positive or negative returns. Using data from the Taiwan Economic Journal, the authors focused on predicting the electronic industry using only fundamental indicators as input. Taking into account average prediction accuracy, type I and II errors and return on investment of these models, they concluded that multiple classifiers outperform single classifiers. In addition, it was also concluded that heterogeneous classifier ensembles performed slightly better than homogeneous classifier ensembles and, finally, there is no significant difference in accuracy prediction between majority voting and bagging methods. The best bagging and voting ensemble models both acquired accuracies of 66%, while accuracies of 63%, 60% and 59% were obtained for MLP-NN, LR and CART respectively.

---

[8]An estimate where the price of a given currency at day t_i is the average price of the same currency between $t_i - w$ and $t_i - 1$ included, w is the window size.

Table 2.1 Summary of the most relevant works covered in the state-of-the-art

Ref.	Year	Financial market	Used methodologies	Dataset time period (data frequency)	Evaluation function	System performance	B&H performance
[61]	2017	Credit default swaps market	GA	1/12/2007–1/12/2016 (daily data)	ROI	87.84% (ROI)	NA
[64]	2016	12 most-volumed cryptocurrencies exchange data	Model-less convolutional NN	27/08/2015–27/08/2016 (30-min data)	Portfolio value maximization	16.3% (ROI portfolio value)	0.876% (ROI portfolio value)
[66]	2018	Bitcoin/USD exchange data	ANN (only w/long positions implemented)	31/07/2016–24/01/2018 (15-min data)	ROI	6.68% (ROI for average constant spread)	2.28% (ROI for average constant spread)
[68]	2016	Bitcoin/USD exchange data	LSTM network	19/08/2013–19/07/2016 (daily data)	Accuracy, RMSE	52.7% (accuracy)	NA
[70]	2015	Bitcoin/USD exchange data	SVM, LR, NN	01/02/2012–01/04/2013 (hourly data)	Accuracy	53.7% for SVM, 54.3% for LR and 55.1% for NN (accuracy)	NA
[77]	2015	Bitcoin/USD exchange data	BOX-SVM, VW-SVM	09/01/2015–02/02/2015 (15-min data)	Accuracy	10.58% for BOX-SVM, 33.52% for VW-SVM (ROI)	4.86% (ROI)
[78]	2019	Bitcoin/USD exchange data	SVM and ensemble (RNN and tree classifier)	01/04/2013–01/04/2017 (daily data)	Accuracy and mean square error	59.4% and 62.9% (accuracy)	NA
[81]	2018	12 most liquid cryptocurrencies exchange data	RF (best performing model)	10/8/2017–23/6/2018 (15-min data)	Accuracy	53% (accuracy)	NA
[85]	2011	Electronic industry from the Taiwan economic journal	MLP, CART, LR, voting and bagging methods (of the 3 single classifiers)	2005 Q4–2006 Q3 (quarterly fundamental data)	Accuracy	Voting methods 4837%, bagging methods 4637% (ROI)	2970% (ROI)

2.4 Conclusions

This chapter starts with a description of the fundamental concepts required to comprehend the system developed in this work. Namely, an overview over the cryptocurrency world and the main characteristics of its exchange market are presented followed by two forms of financial data analysis. After that, all financial indicators applied in this work are introduced.

Secondly, the basic concepts of the machine learning algorithms utilized in this work are briefly introduced. It should be pointed out that each individual classifier algorithm has much more to be said, which cannot be covered by this work. In order to obtain a more comprehensive understanding of this work's employed machine learning algorithms, the literature referenced throughout this section contains further and more elaborate material.

Finally, the most relevant research conducted for the creation of this work is presented. Besides including a general outline and conclusions, applicable details or considerations that were taken under consideration when designing this work are also described. Out of the several mentioned works, the most pertinent ones are summarized in Table 2.1. All things considered, the analysis carried out over literature had a crucial contribution towards this work and provided the starting point to accomplish this work's objectives.

References

1. Chaum D (1983) Blind signatures for untraceable payments. Advances in cryptology. Springer, Boston, pp 199–203
2. Arner DW (2009) The global credit crisis of 2008: causes and consequences. Int Lawyer 43:91
3. Penrose KL (2013) Banking on Bitcoin: applying anti-money laundering and money transmitter laws. NC Bank Inst 18:529
4. Gurría A (2009) Responding to the global economic crisis. OECD's role in promoting open markets and job creation. Accessed 30 Jan 2018
5. Acharya V, Philippon T, Richardson M, Roubini N (2009) The financial crisis of 2007-2009: causes and remedies, New York University Salomon Center and Wiley Periodicals
6. Nakamoto S (2008) Bitcoin: a peer-to-peer electronic cash system
7. Bonneau J, Miller A, Clark J, Narayanan A, Kroll JA, Felten EW (2015) SoK: research perspectives and challenges for bitcoin and cryptocurrencies. In: 2015 IEEE symposium on security and privacy. IEEE, pp 104–121
8. Antonopoulos AM (2014) Mastering Bitcoin: unlocking digital cryptocurrencies. O'Reilly Media Inc., Sebastopol
9. White LH (2015) The market for cryptocurrencies. Cato J 35:383
10. Dierksmeier C, Seele P (2018) Cryptocurrencies and business ethics. J Bus Ethics 152(1):1–14
11. Yermack D (2015) Is Bitcoin a real currency? An economic appraisal. Handbook of digital currency. Academic, New York, pp 31–43
12. Gronwald M (2014) The economics of Bitcoins–market characteristics and price jumps
13. Dyhrberg AH (2016) Bitcoin, gold and the dollar-A GARCH volatility analysis. Financ Res Lett 16:85–92
14. Glaser F, Zimmermann K, Haferkorn M, Weber MC, Siering M (2014) Bitcoin-asset or currency? Revealing users' hidden intentions (15 April 2014). ECIS

15. Dwyer GP (2015) The economics of Bitcoin and similar private digital currencies. J Financ Stab 17:81–91
16. Chan S, Chu J, Nadarajah S, Osterrieder J (2017) A statistical analysis of cryptocurrencies. J Risk Financ Manag 10(2):12
17. Cheung A, Roca E, Su JJ (2015) Crypto-currency bubbles: an application of the Phillips-Shi-Yu (2013) methodology on Mt. Gox bitcoin prices. Appl Econ 47(23):2348–2358
18. Chuen DLK (ed) (2015) Handbook of digital currency: Bitcoin, innovation, financial instruments, and big data. Academic, New York
19. Gandal N, Halaburda H (2014) Competition in the cryptocurrency market
20. CoinMarketCap. Cryptocurrency exchange rankings. https://coinmarketcap.com/rankings/exchanges/. Accessed 02 Sept 2020
21. Murphy JJ (1999) Technical analysis of the financial markets: a comprehensive guide to trading methods and applications. Penguin
22. Fama EF (1995) Random walks in stock market prices. Financ Anal J 51(1):75–80
23. Achelis SB (2001) Technical analysis from A to Z
24. De Prado ML (2018) Advances in financial machine learning. Wiley, New York
25. Fernández-Blanco P, Bodas-Sagi DJ, Soltero FJ, Hidalgo JI (2008) Technical market indicators optimization using evolutionary algorithms. In: Proceedings of the 10th annual conference companion on genetic and evolutionary computation, pp 1851–1858
26. Cavalcante RC, Brasileiro RC, Souza VL, Nobrega JP, Oliveira AL (2016) Computational intelligence and financial markets: a survey and future directions. Expert Syst Appl 55:194–211
27. Abu-Mostafa YS, Atiya AF (1996) Introduction to financial forecasting. Appl Intell 6(3):205–213
28. Lo AW, Mamaysky H, Wang J (2000) Foundations of technical analysis: computational algorithms, statistical inference, and empirical implementation. J Financ 55(4):1705–1765
29. Teixeira LA, De Oliveira ALI (2010) A method for automatic stock trading combining technical analysis and nearest neighbor classification. Expert Syst Appl 37(10):6885–6890
30. Kirkpatrick CD II, Dahlquist JA (2010) Technical analysis: the complete resource for financial market technicians. FT Press, Upper Saddle River
31. Guresen E, Kayakutlu G, Daim TU (2011) Using artificial neural network models in stock market index prediction. Expert Syst Appl 38(8):10389–10397
32. McNelis PD (2005) Neural networks in finance: gaining predictive edge in the market. Academic, New York
33. Hosmer DW Jr, Lemeshow S, Sturdivant RX (2013) Applied logistic regression, vol 398. Wiley, New York
34. Raschka S (2015) Python machine learning. Packt Publishing Ltd., Birmingham
35. Murphy KP (2012) Machine learning: a probabilistic perspective. MIT Press, Cambridge
36. Breiman L (2001) Random forests. Mach Learn 45(1):5–32
37. Breiman L (1996) Bagging predictors. Mach Learn 24(2):123–140
38. Venables WN, Ripley BD (2013) Modern applied statistics with S-PLUS. Springer Science & Business Media, New York
39. Hastie T, Tibshirani R, Friedman J (2009) The elements of statistical learning: data mining, inference, and prediction. Springer Science & Business Media, New York
40. Probst P, Wright MN, Boulesteix AL (2019) Hyperparameters and tuning strategies for random forest. Wiley Interdiscip Rev: Data Min Knowl Discov 9(3):e1301
41. Louppe G (2014) Understanding random forests: from theory to practice. arXiv:1407.7502
42. Chen T, Guestrin C (2016) XGBoost: a scalable tree boosting system. In: Proceedings of the 22nd ACM SIGKDD international conference on knowledge discovery and data mining, pp 785–794
43. Freund Y, Schapire RE (1996) Experiments with a new boosting algorithm. In: ICML, vol 96, pp 148–156
44. Freund Y, Schapire RE (1997) A decision-theoretic generalization of on-line learning and an application to boosting. J Comput Syst Sci 55(1):119–139

45. Schapire RE, Freund Y, Bartlett P, Lee WS (1998) Boosting the margin: a new explanation for the effectiveness of voting methods. Ann Stat 26(5):1651–1686
46. Friedman JH (2002) Stochastic gradient boosting. Comput Stat Data Anal 38(4):367–378
47. Friedman J, Hastie T, Tibshirani R (2000) Additive logistic regression: a statistical view of boosting (with discussion and a rejoinder by the authors). Ann Stat 28(2):337–407
48. Friedman JH (2001) Greedy function approximation: a gradient boosting machine. Ann Stat 1189–1232
49. Mitchell R, Frank E (2017) Accelerating the XGBoost algorithm using GPU computing. PeerJ Comput Sci 3:e127
50. Cortes C, Vapnik V (1995) Support-vector networks. Mach Learn 20(3):273–297
51. Fletcher T (2009) Support vector machines explained. Tutorial paper
52. Platt J (1999) Probabilistic outputs for support vector machines and comparisons to regularized likelihood methods. Adv Large Margin Classif 10(3):61–74
53. Lin HT, Lin CJ, Weng RC (2007) A note on Platt's probabilistic outputs for support vector machines. Mach Learn 68(3):267–276
54. James G, Witten D, Hastie T, Tibshirani R (2013) An introduction to statistical learning, vol 112. Springer, New York, p 18
55. Dietterich TG (2000) Ensemble methods in machine learning. In: International workshop on multiple classifier systems. Springer, Berlin, pp 1–15
56. Zhou ZH (2012) Ensemble methods: foundations and algorithms. CRC Press, Boca Raton
57. Webb GI, Zheng Z (2004) Multistrategy ensemble learning: reducing error by combining ensemble learning techniques. IEEE Trans Knowl Data Eng 16(8):980–991
58. Brockwell PJ, Davis RA, Calder MV (2002) Introduction to time series and forecasting, vol 2. Springer, New York, pp 3118–3121
59. Tsay RS (2005) Analysis of financial time series, vol 543. Wiley, New York
60. Box GE, Jenkins GM, Reinsel GC (2011) Time series analysis: forecasting and control, vol 734. Wiley, New York
61. Cardoso JR, Neves R (2017) Investing in credit default swaps using technical analysis optimized by genetic algorithms. Instituto Superior Tecnico
62. Abreu G, Neves R, Horta N (2018) Currency exchange prediction using machine learning, genetic algorithms and technical analysis. arXiv:1805.11232
63. Shah D, Zhang K (2014) Bayesian regression and Bitcoin. In: 2014 52nd annual Allerton conference on communication, control, and computing (Allerton). IEEE, pp 409–414
64. Jiang Z, Liang J (2017) Cryptocurrency portfolio management with deep reinforcement learning. In: 2017 intelligent systems conference (IntelliSys). IEEE, pp 905–913
65. Nadkarni J, Neves RF (2018) Combining NeuroEvolution and principal component analysis to trade in the financial markets. Expert Syst Appl 103:184–195
66. Nakano M, Takahashi A, Takahashi S (2018) Bitcoin technical trading with artificial neural network. Phys A: Stat Mech Appl 510:587–609
67. Yao J, Tan CL (2000) A case study on using neural networks to perform technical forecasting of forex. Neurocomputing 34(1–4):79–98
68. McNally S, Roche J, Caton S (2018) Predicting the price of Bitcoin using machine learning. In: 2018 26th Euromicro international conference on parallel, distributed and network-based processing (PDP). IEEE, pp 339–343
69. Atsalakis GS, Valavanis KP (2009) Surveying stock market forecasting techniques-part II: soft computing methods. Expert Syst Appl 36(3):5932–5941
70. Greaves A, Au B (2015) Using the Bitcoin transaction graph to predict the price of Bitcoin. No Data
71. Colianni S, Rosales S, Signorotti M (2015) Algorithmic trading of cryptocurrency based on Twitter sentiment analysis. CS229 Project, 1-5
72. Stenqvist E, Lönnö J (2017) Predicting Bitcoin price fluctuation with Twitter sentiment analysis
73. Ballings M, Van den Poel D, Hespeels N, Gryp R (2015) Evaluating multiple classifiers for stock price direction prediction. Expert Syst Appl 42(20):7046–7056

74. Ou P, Wang H (2009) Prediction of stock market index movement by ten data mining techniques. Mod Appl Sci 3(12):28–42
75. Kumar M, Thenmozhi M (2006) Forecasting stock index movement: a comparison of support vector machines and random forest. In: Indian institute of capital markets 9th capital markets conference paper
76. Madan I, Saluja S, Zhao A (2015) Automated Bitcoin trading via machine learning algorithms 20
77. Żbikowski K (2015) Using volume weighted support vector machines with walk forward testing and feature selection for the purpose of creating stock trading strategy. Expert Syst Appl 42(4):1797–1805
78. Mallqui DC, Fernandes RA (2019) Predicting the direction, maximum, minimum and closing prices of daily Bitcoin exchange rate using machine learning techniques. Appl Soft Comput 75:596–606
79. Huang W, Nakamori Y, Wang SY (2005) Forecasting stock market movement direction with support vector machine. Comput Oper Res 32(10):2513–2522
80. Kim KJ (2003) Financial time series forecasting using support vector machines. Neurocomputing 55(1–2):307–319
81. Akyildirim E, Goncu A, Sensoy A (2020) Prediction of cryptocurrency returns using machine learning. Ann Oper Res 1–34
82. Basak S, Kar S, Saha S, Khaidem L, Dey SR (2019) Predicting the direction of stock market prices using tree-based classifiers. N Am J Econ Financ 47:552–567
83. Alessandretti L, ElBahrawy A, Aiello LM, Baronchelli A (2018) Anticipating cryptocurrency prices using machine learning. Complexity 2018
84. do Ó Barbosa R, Neves R (2018) Ensemble of machine learning algorithms for economic recession detection. Instituto Superior Tecnico
85. Tsai CF, Lin YC, Yen DC, Chen YM (2011) Predicting stock returns by classifier ensembles. Appl Soft Comput 11(2):2452–2459

Chapter 3
Implementation

3.1 System's Architecture

First and foremost, it is worth noting that no single model is obviously or consistently better than all others, each model has trade-offs. The "No Free Lunch" theorem [1] states that any two optimization algorithms are equivalent when their performance is averaged across all possible problems. However, for this thesis, a particular problem is being solved, thus some algorithms will in fact perform better than others.

In this thesis a trading system containing 5 different methodologies for forecasting is used to detect the best entry and exit points in the cryptocurrency market. In order to predict these best entry and exit points in a financial market, the direction of price, rather than price levels, is forecast in this work. This method has proven to be effective and profitable by many authors in literature as can be seen in Chap. 2. Simply put, this work attempts to solve a binary classification problem. The cryptocurrency market is known for its volatility, as so, this proposed system is intended to take advantage of the intense and rapid variations through several mechanisms. The main process to tackle this issue is the resampling of financial data to a parameter other than time. Through resampling, periods of time with large variations are given a larger importance relatively to tranquil periods of time. Additional procedures were added, such as stop-loss orders to curb losses or line of action chosen for the target formulation. These processes are explained in further detail throughout this chapter.

It is expected beforehand that the predictions made from the ensemble will live up to expectations by exceeding the performance made from each individual learner's predictions. In order to create a good ensemble, it is generally believed that the base learners should be as accurate and diverse as possible [2]. Thus, in this proposed system a set of individual learners with these characteristics was chosen.

LR is one of the most widely used linear statistical model in binary classification situations [3]. It is a simple and easily implementable model that offers a reasonable performance, as was seen in Sect. 2.3. The linear SVM and non-linear RF methods, are used due to their well above average performances. Throughout supervised

© The Author(s), under exclusive license to Springer Nature Switzerland AG 2021

T. A. Borges and R. Neves, *Financial Data Resampling for Machine Learning Based Trading*, SpringerBriefs in Computational Intelligence, https://doi.org/10.1007/978-3-030-68379-5_3

learning literature [4] and as was seen in Sect. 2.3, when compared to other learning methods, these two methods generally achieve top performances and are recommended by the authors. GTB is a non-linear method employed through the extreme gradient boosting (XGBoost) framework. This method is an efficient and scalable implementation of gradient boosting known for winning many machine learning competitions. In fact, in 2015 it was the winning algorithm for 17 out of 29 *Kaggle*[1] challenges [5].

Furthermore, these four methods provide the coefficient size for each feature, which can be translated into how relevant each feature is. LR and SVM additionally provide a measure of the direction (positive or negative) of the association of each feature. Accordingly, superfluous features could be withdrawn in an attempt to fine-tune the utilized set of features. This was considered as an advantage over Artificial Neural Network related models where due to its multi-layered process, an idea of the relationship between inputs and outputs is not provided [6]. These methods, commonly referred to as *black box* methods, are powerful in pure predictive settings, such as pattern recognition, but are less useful in data mining processes, where only a fraction of the large number of predictor variables included are, in fact, relevant to the prediction [6].

For this system, the common systematic procedure to predict time series using machine learning algorithms was put into practice by following these steps in this specific order [7]: data preparation, algorithm definition, training and finally forecasting evaluation. With this purpose, the proposed system is divided into 4 modules as briefly follows:

- Data module (Sect. 3.2): This module is responsible for collecting the data from Binance's API and preparing it to be used by the next module. The data preparation consists in selecting the pairs with most volume, erasing unessential information and lastly resampling the pairs through four different methods.
- Technical Rules module (Sect. 3.3): This module receives the processed data from the data module and calculates a set of technical indicators for each data point to be posteriorly used as input for the following module.
- Machine Learning module (Sect. 3.4): This module is the main core of the system. Having the data rearranged and the technical indicators calculated, this module is responsible for splitting the data into train and test datasets, doing another round of data preparation (standardization), training the different machine learning methodologies for all the training intervals and finally generating forecasts using the trained predictive models for all the test intervals. These forecasts, referred to as "trading signals", are the output sent to the next module.
- Investment Simulator module (Sect. 3.5): Each trading signal is now tested in a real world situation. A number of statistical indicators such as *accuracy*, as well as financial indicators such as *return on investment* are calculated, plotted and stored in memory.

---

[1]Online community of data scientists and machine learners known for hosting hundreds of machine learning competitions since its inception.

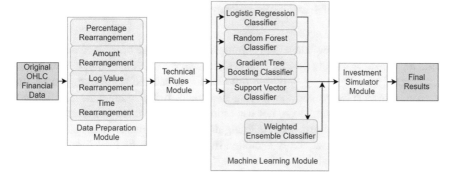

Fig. 3.1 Simplified overview of the proposed system's modular architecture

This modular approach was adopted to provide independence, flexibility and understandability to each of the modules. Consequently, any sort of extension or modification to any of the modules is facilitated. In Fig. 3.1 the main phases of the system as well as the interactions between each module are shown.

3.2 Data Preparation Module

The starting point for this work's proposed system is a collection of homogeneously sampled time series that will be processed in this module in order to acquire informative features. The original time series were all retrieved through Binance's API.[2] For each cryptocurrency exchange pair, the obtained time series contained data for a specified time frame and a specified time frequency. In this thesis the samples used are as detailed as one can get from Binance: 1 sample per minute.

In this system each data point utilizes the customary fields required to build an OHLC (open-high-low-close) chart:

- Open time: Starting time for a given trading time period with the format "YYYY-MM-DD hh:mm:ss".
- Open: Price in which a cryptocurrency pair is first traded upon the opening of a given trading period.
- High: Highest price in which a cryptocurrency pair was traded for a given trading period.
- Low: Lowest price in which a cryptocurrency pair was traded for a given trading period.
- Close: Price in which a cryptocurrency pair is last traded upon the opening of a given trading period. It's worth pointing out that in this work's system, it is assumed that trading is contracted by these closing prices.

---

[2]The python package "Binance official API" (https://github.com/binance-exchange/binance-official-api-docs/) was utilized to retrieve the used historical data from Binance's API (https://api.binance.com)

- Volume: Amount of base cryptocurrency traded during the respective trading period.

Binance offered a few more data columns, such as 'close time' indicating ending time for the same trading period. However, as every single data point contains data for precisely 1 min, using only one time label per sample is enough. Hence, the open time was selected and the closing time field, being redundant, was deleted. A few more data columns were not used throughout the whole system so, in order to spare computer memory, the remaining downloaded columns were simply freed from memory and consequently excluded from storage in the final file. In the end, the data columns containing each of these 6 fields, the only data remaining, were saved into a CSV (comma-separated values) file.

In this work the several alternative bar representations attained through data resampling are to be tested and compared to the common time sampling, thus, the previously saved dataset containing the Open time, Open, High, Low, Close and Volume columns is now to be resampled. As mentioned in Sect. 2.3.1, contrarily to the customary time representation, resampled data is intended to place more importance in high-frequency intervals by overrepresenting the constituting individual samples, when compared to low-frequency intervals. In addition two more important aspects of utilizing resampled data should be noted:

1. From a machine learning perspective, if both high and low-frequency intervals are equally represented, a mistaken prediction would yield an equivalent penalization for both cases. Now, assuming high-frequency intervals are overrepresented relatively to low-frequency intervals, if a high-frequency interval with many samples (all consisting of small time intervals) were to be erroneously predicted, a penalization for each single misclassification would be appointed resulting in a collectively large penalization. On the other hand, an erroneous prediction on a low-frequency interval would also be penalized, but not as heavily since it contains less samples to classify when compared to a high-frequency interval. To put it simply, the penalizations for incorrect predictions can be exploited in order to favour a better performance on high-frequency intervals rather than in low-frequency intervals throughout the whole forecast;
2. From a financial perspective, it is more important to successfully forecast high-frequency periods as larger returns or losses can be obtained when compared to the more stable low-frequency periods.

In this work four different resampling processes are applied. All of them follow the same formula, thus, only the first will be described in full detail. The explanation of the remaining three methods are abbreviated and contrasted to the first resampling method. A description of each grouping now follows:

- **Percentage**: The purpose of this type of rearrangement is to sample data according to a fixed percentual variation threshold rather than to a fixed time period. With this intent, the first step is to calculate the percentual variation between the consecutive samples of the original time sampled dataset, i.e., the percentual first order differencing of the closing values. As was mentioned in Sect. 2.3.1, the first order differencing of a time series consists on calculating the difference between a given value and the consecutive previous value. In this case, the percentual first order differencing of close values is calculated: the percentual difference from the close of an actual data point with its immediate previous data point are stored on a column with indexes synchronized with the actual data point.

Knowing the absolute variation of each sample, the objective is now to identify sets of consecutive samples whose individual variations aggregated together reach or exceed a fixed pre-defined threshold of absolute percentual variation. For this purpose, a new column was created to contain the result of a customized cumulative sum of the previously created column (percentual first order differencing of close values). This customized cumulative sum is responsible for defining the boundaries of each group through assigning different numerical identifiers to each sample. To do so, starting on a specific sample, the total cumulative absolute variation of the specific sample as well as the consecutively posterior samples are added up until the variation sum reaches or exceeds the fixed threshold. This occurrence dictates the starting and ending boundary for each group: whenever a threshold is crossed, a group ends on the actual sample and a new one begins on the next sample, with the next numerical identifier. In the next group, the cumulative sum is restarted back to zero and the same iterative method is followed, until the cumulative absolute variation summed reaches or exceeds the threshold, all samples are to be grouped together.

In the end, all samples of the original dataset with the same identifiers are grouped into a single sample of the final dataset. That is, each original sample is attributed a group identifier in accordance with the modified cumulative sum. This process is done in a orderly manner that iterates through all samples (in a descendant order, where the oldest entry is at the top and the most recent is at the bottom), therefore, only consecutive samples can be grouped together. This process is illustrated in Fig. 3.2. The seventh column (Cumulative sum w/restart when threshold is reached) contains the values obtained with the customized cumulative sum algorithm. As can be observed, whenever the value of this seventh column is equal or larger than a fixed threshold, in this case 2%, the group being numbered with the identifier 'i' is closed, the value for the cumulative sum is restarted and a new group with identifier '$i + 1$' is initiated in the next sample. In the end of this iterative process each individual sample of the original dataset contains a numerical group identifier associated to it.

This modified cumulative sum algorithmic formula, is responsible for resampling the original dataset into one whose sample's size is as close to the fixed threshold value as possible resulting in an almost even sized (in this case with the same percentual size) dataset. Furthermore, leaking of future data is prevented, as new groups are defined only once the threshold is reached or surpassed. Through this method, in the end, the same amount of percentual closing price action, is approximately the same for each sample in the resampled dataset. It's worth adding that financial markets generate a discrete time series, making it impossible to achieve a perfectly even-sized resampled dataset with data from any cryptocurrency market. Nevertheless, a finer-grained periodic (smallest possible time period between data points) original data, independently of the threshold value, is more likely to build a more regularly and consistently sized resampled dataset. Similarly to utilizing finer-grained data, if the threshold is increased, uneven resampled data becomes less noticeable as differences between resampled data points become percentually more insignificant resulting in a seemingly more consistent dataset. The values considered for the threshold should be larger than the payed fees, so each profitable transaction covers the fees completely, but not too large, otherwise too many data points would be grouped resulting in loss of information and consequently profit loss.

Lastly, as previously mentioned, all data points containing the same numerical group identifier must be grouped into a single data point. To accomplish this, the Open and Close values of the new resampled data point are respectively the open value of the first entry and the close value of the last entry in the set of samples that make part of the respective group. The new High value is the highest of high values out of all the entries in the group and the new Low value is the lowest. Lastly, the new Volume corresponds to the sum of all volumes for each data point in the group. In case a group consists of only one data point, meaning a percentual variation larger than the chosen threshold happened in that instant, one of two cases happens: either the actual group is closed (e.g. row 11 in the original dataset of Fig. 3.2); or, if this new row is supposed to initiate a new group, a group is started and closed in the same instant (e.g. row 12 in the original dataset of Fig. 3.2). For the second case, there is no need to rearrange that data point, as it is already in the final form. One last detail worth mentioning is that, grouping is done by comparing two consecutive entries, the initial data point of the original dataset is only used as reference, it does not contribute towards any group. However, the impact caused by discarding this single sample is negligible as the original dataset for each cryptocurrency pair on average contains hundreds of thousands of data points.

- **Amount**: The purpose of this type of rearrangement is to sample data according to a fixed variation threshold rather than to a fixed percentage or time period. This process is identical to the Percentage rearrangement but is even simpler. Instead of calculating the percentual first order difference between closes, simply the actual first order difference series of the closing values is calculated. As was mentioned in Sect. 2.3.1, the first order difference of the close series corresponds to the changes from the close of an actual data point with its immediate previous data point and has the added benefit of achieving (or approaching) stationarity.

  All procedures for calculating the alternative cumulative sum column, for assignment of new group identifiers and sample grouping are identically followed throughout the remainder of this resampling process. The only exception is the fixed threshold and how it is determined. In this case, a fixed price value is used as threshold instead. This fixed value corresponds to a fixed closing price variation of the base currency expressed in units of quote currency. To obtain this fixed value, firstly, the sum of absolute differences between closes is calculated, i.e., $Sum(abs(differencing(closeprices)))$. The value obtained is divided by the total number of data points obtained in the percentage grouping. This proce-

	Open	High	Low	Close	Close absolute percentual variation	Cumulative sum w/restart when 2% threshold is exceeded	Newly assigned group number
0	4.883	4.883	4.883	4.883	NA	NA	NA
1	4.856	4.954	4.856	4.954	1.45%	1.45%	0
2	4.948	4.948	4.945	4.945	0.18%	1.64%	0
3	4.935	4.935	4.857	4.857	1.78%	3.42%	0
4	4.856	4.916	4.856	4.915	1.19%	1.19%	1
5	4.915	4.915	4.856	4.903	0.24%	1.44%	1
6	4.902	4.902	4.9	4.9	0.06%	1.50%	1
7	4.9	4.955	4.9	4.954	1.10%	2.60%	1
8	4.954	4.954	4.954	4.954	0%	0%	2
9	4.955	5.067	4.864	5.062	2.18%	2.18%	2
10	5.058	5.123	5.058	5.121	1.17%	1.17%	3
11	5.12	5.12	4.981	4.981	2.73%	3.90%	3
12	4.95	5.123	4.94	5.123	2.85%	2.85%	4

	Open	High	Low	Close
-	4.883	4.883	4.883	4.883
0	4.856	4.954	4.856	4.857
1	4.856	4.955	4.856	4.954
2	4.954	5.067	4.864	5.062
3	5.058	5.123	4.981	4.981
4	4.95	5.123	4.94	5.123

(a) Table on the left contains the original dataset that is resampled into the dataset on the right.

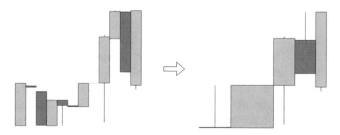

(b) Left candlestick group is a representation of the original dataset and right candlestick group is a representation of the resampled dataset.

Fig. 3.2 Regrouping process illustration for a percentage of 2%

dure guarantees that the quantity of data points obtained is similar to the quantity obtained in percentage grouping. Nonetheless it is unlikely that the final amount of samples obtained correspond precisely to the amount of samples obtained in the first rearrangement. These two types of resampling procedures group samples according to different variations which may phase themselves differently resulting in less groups during the customized cumulative sum group identifier assignment process and consequently less samples in the final dataset. Anyhow, this is not a frequent occurrence and the total sample difference between this rearrangement and the percentage rearrangement is never substantial (it is never larger than 1% of the total amount of samples of each rearrangement). In the end, likewise to the percentage rearrangement, the same amount of closing price action, is approximately the identical for each candlestick.

- **Logarithmic Amount**: This process can be explained as an extension to the amount rearrangement. The only distinction is that a natural logarithm is applied to the closing prices before calculating the first difference series of the closing values: the difference between each data point and the consecutive previous one are calculated in a logarithmic scale.

Once again, apart from using a different fixed value to group samples, all other processes are conducted identically to the previous procedures. This rearrangement's fixed value is calculated analogously to the procedure used in the previous rearrangement but with the logarithmic scale taken into account: the final sum of absolute differences between each consecutive close value in a natural logarithmic scale, i.e., $Sum(abs(differencing(log(closingprices))))$, is divided by the total number of data points obtained in the percentage grouping. Once more, this was chosen so that the total quantity of rearranged data points for this process is similar to the total quantity of data points obtained in the two previous groupings. Subsequently, the customized cumulative sum column is calculated, each sample is assigned to a group and actual resampling happens identically to what was described in the percentage resampling method. In the end, likewise the two previous rearrangements, the same amount of closing price action in a logarithmic scale, is approximately the same for each candlestick.

The purpose of this resampling process is similar to the simple first differencing of a time series utilized in the amount rearrangement case and introduced in Sect. 2.3.1. However, logarithmic scaling makes better use of input data scope by evenly spreading data across the input range. It is a simple but effective way to stabilize the variance across time. In most cases, as variation in volatility is larger for the series without logarithms, using a logarithmic scale is a means for stabilizing the variance [8]. However, even the first differences of logarithms still show considerable variation in the volatility. While the difference series (amount resampling) itself may still display a non-stationary behaviour, a log transformation with differencing is more likely to ensure a stationary condition [9]. Furthermore, the logarithm converts multiplicative relationships to additive relationships, and exponential trends to linear trends. The logarithmic transformation straightens out exponential growth patterns and stabilizes variance. The logarithmic resampling is more amenable to statistical analysis than the simple amount resampling because it is easier to derive the time series properties of additive processes than those of multiplicative processes [10].

Changes (up to $\pm 15\%$) in the natural log of a variable can be directly interpretable as percent changes [11]. Nevertheless, contrarily to percentages, a decrease in a certain differenced log value followed by a same-sized increase takes the value to the same position, whereas a given percentual loss followed by the same percentual gain ends up in a worse position. It follows that the slope of a trend line of the logged data is similar to the average percentage growth, as changes are similar to percentage changes [12]. In conclusion, logarithmic resampling has potential to perform better than the two other resampling methods.

- **Time**: This resampling variety was implemented with the aim of being the comparison baseline as it is by far the most commonly used layout for sampling financial time series in literature [13].

In spite of the original dataset being already sampled according to time, due to the lack of rearrangement, it contains plenty more data points than each resulting re-sampled dataset of the three previous rearrangements. This would most certainly generate unfair results. A larger dataset could generate much larger profits, or losses. Therefore, to make comparisons between the different resampling methods fair, a time grouping must also be implemented. This type of resampling consists on simply grouping together a fixed amount of consecutive data points. The fixed amount of consecutive data points to be aggregated is obtained identically to the previous rearrangements: calculate the integer division between the total amount of data points and the average amount of rearranged data points of the three previous rearrangements.

In the end contrarily to the three previous rearrangements, candlesticks are not grouped according to an amount of closing variation but according to a fixed interval of time as is customary. It's worth mentioning that despite the final total number of samples rarely coinciding with the previous 3 groupings, similarly to the differences between the amount and percentage rearrangements, the total difference between the number of samples of each rearrangement is never significant whatsoever.

The procedures utilized in this module were fully optimized and made use of parallelization methods with the intention of reducing this module's total execution time. In the end, this module turned out relatively fast, seldom taking over 2 s even for large datasets, therefore, there was no need to store the output in memory.

3.3   Technical Indicators Module

The purpose of this proposed system is predicting future values of financial markets, based on technical indicators. This module is responsible for generating and outputting the respective technical indicators for each data sample. To accomplish this, the rearranged time series are received as input and several technical indicators are calculated for each data sample. The input dataset corresponds to the output of the Data Module.

Using technical analysis simplifies the forecasting future market movements problem to a pattern recognition problem, where inputs are derived from historical prices and outputs are an estimate of the price or its trend [14].

In literature, researchers have used different types of technical indicators to monitor the future movements of prices and in setting up trading rules. In this proposed system, the used indicators were chosen by gathering a set of the most commonly used technical indicators in machine learning algorithms destined to forecast financial markets. Each indicator reveals information about a certain characteristic of the said financial market, hence, using a diversified set of technical indicators was given importance. The used set of technical indicators and the values for each respective parameter are shown in Table 3.1. All technical indicators, except for one additional variant utilized in the Stochastics, are explained in Sect. 2.2.1. It's worth mentioning that relatively to the SMA, the EMA reacts faster to price variations yielding better results in forecasting the cryptocurrency market, hence only the latter was utilized.

Table 3.1 List of all indicators and (if applied) their respective parameters, outputted to the machine learning module

Technical indicator	Parameters
EMAs	5, 10, 20, 50, 100, 200 periods
Relative strength index	14 periods
Moving average convergence divergence histogram	Signal EMA 9, fast EMA 12, slow EMA 26 periods
On balance volume	–
Rate of change	10 periods
Commodity channel index	14 periods
Average true range	14 periods
Stochastic oscillator %K line	14 periods
Stochastic oscillator fast %D line	3 periods
Stochastic oscillator slow %D line	3 periods
Histogram between %K line & fast %D line	–
Histogram between %K line & slow %D line	–

The utilized technical indicator variant, consists on replicating the histogram method from the MACD indicator on the stochastic oscillators to indicate trend reversals (when a sample of this indicator crosses zero) as well as whether the trend is upwards or downwards. For this purpose, two histograms were used: the difference between %K line and slow %K line is the first, and the difference between %K line and fast %D line is the second. Aside from this variant all the technical indicators used in this thesis were implemented as described in Sect. 2.2.1. Table 3.1 lists the used technical indicators as well as the values for each parameter (if existent). In this proposed system, the values used for each parameter are the ones traditionally used or suggested.

The dataset containing the calculated technical indicators in this module is outputted to the next module where it will be responsible for generating the input features used to train the four binary classifiers. This module is relatively fast, hence, there was no need to store the output in memory.

3.4   Machine Learning Module

The output data from the previous module, technical indicator module, is received by this module. At this point, the useful fields of financial data have been picked out, grouped and the technical indicators for each data point have been calculated. This module is the most complex and the main core of the system, as so it was divided into 3 main components, each with a different responsibility.

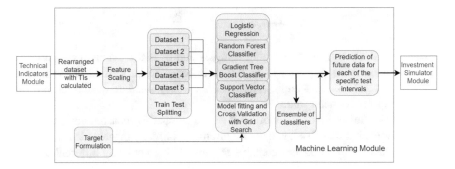

Fig. 3.3 Machine learning module overview

The first component is responsible for scaling the features (the input data containing the technical indicators) and posteriorly the best features are selected to be used. Feature scaling is a actually a requirement for some of the machine learning algorithms employed in this proposed system.

The second step undertaken in this module is defining the target prediction vector. This component is necessary because all methods of binary classification used in this system are a type of supervised learning, which by definition require a target prediction.

Lastly, only when both the target prediction vector is defined (also known as *vector y*) and the financial data is in its final form and ready to be used as the sample matrix (also known as *matrix X*), is the system ready for the third and last component. The aim of this third component is fitting each of the 4 binary classifiers, generating a forecast for each algorithm and in the end creating an average of the forecasts from the four algorithms making an ensemble of machine learning algorithms, the fifth prediction signal of this system. All 5 predictions are outputted to the Investment Simulator module, the last module of this system, to be translated into a trading signal and simulated into a real market environment.

Figure 3.3 contains a brief illustration of this module's main steps.

3.4.1 Feature Scaling

Feature scaling is a data preprocessing step, applied in this thesis through a technique called *standardization*, represented in Eq. (3.1). Standardization is the process of centring features by subtracting the features' mean and scaling by dividing the result by the features' standard deviation. Standardization centres the features at mean 0 with standard deviation 1, so that the range of each independent feature has the properties of a standard normal distribution. The standardized value, also known as *z-score* or *standard score*, of a given sample, x, is calculated as follows:

$$x_{standardized} = \frac{x - \mu}{\sigma}, \qquad (3.1)$$

where μ is the mean of the training samples and σ is the standard deviation of the training samples. Equation (3.1), which is based on properties of the normal distribution rescales each sample in terms of how many standard deviations they fall above or below the mean value.

Feature scaling is not only important if we are comparing measurements that have different units, but it is also a general requirement for many machine learning algorithms. When the input dataset contains features with completely different units of measurement, which is the case in this system, using feature scaling avoids the problem in which some features come to dominate solely because they tend to have larger values than others. Processes used in this system, namely by gradient descent, such as, regularization or optimization employed in algorithms such as Logistic Regression or Support Vector Machines, are sensitive to magnitudes and consequently require feature scaling in order to work properly [3, 15].

Tree-based methods are the only family of algorithms who are scale-invariant[3] [6]. Even though Gradient Decision Tree Boosting utilizes gradient descent, similarly to RF, the base learners are trees which are invariant to scaling, hence, feature scaling is inessential for both [3]. However, after verifying that standardized data did not harm the results of the Random Forest or Gradient Decision Tree Boosting classifiers this procedure was used in all algorithms in order to facilitate comparisons between algorithms.

Data standardization happens independently on each feature by computing the relevant statistics on the samples in the training set. The parameters of the standardization from the training dataset are stored in order to standardize the test set or any new unseen data point. Utilizing the stored mean and standard deviation obtained from the training dataset avoids any leaking of future data into the test set, commonly referred to as *look-ahead bias*.[4]

3.4.2 Target Formulation

The objective of the second component of this module is defining the target discrete values to be predicted, the class label, for each data point. This process is a requirement when using methods of supervised learning.

In order to decide whether the signal in the next instant has one of the two possible outcomes, a deterministic binary classification is proposed. In this proposed system, a binary outcome was taken into account: the cryptocurrency market either has a positive or a negative variation in the next instant. In the rare cases where the variation

---

[3]Feature of objects that do not change if variables are multiplied by a common factor, thus represent universality.

[4]Usage of data or information that would have not been available during the period under analysis.

is precisely null, the outcome of the previous instant is duplicated into the current one.

If a signal was previously declining and then stagnated, the instant of stagnation is classified as a negative variation because no profit would be made if an investor entered a long position in this stagnated instant. Equivalently, if a signal was previously increasing and then stagnated, the stagnation instant is classified as a positive variation, because there is no need to leave the market right away and pay trading fees, the increasing trend might persist. This concept is based on the belief that long term trends more often than not will persist rather than reverse. Hence, in the long run the strategy of duplicating the previous outcome will be beneficial relatively to the strategy of attributing a fixed classification (even though this situation seldom occurs due to all the decimal cases).

Before going into further detail regarding this system's target formulation, transaction fees and bid-ask spreads ought to be introduced. The market of cryptocurrencies is a specially volatile market where short term trend reversals are common. Recurrent reversals would result in entering and leaving the market too often, thereby exhausting the initial investment on trading fees. Furthermore, frequent small increases in value, which are bullish by definition, may result in a loss if the trading fees are larger than the increase. Having these two cases considered, the buy and sell trading fees should be taken into account, functioning as a safeguard against these two cases when determining the ideal long position start and ending points. As the financial data was downloaded from Binance, the trading fees applied in this system are Binance's general trading fee of 0.1% [16].

Throughout literature, more often than trading fees, the bid-ask spread is overlooked. While its impact may be negligible on specific cases, it may also be responsible for harming the system's investments in a real situation if unaccounted for. This *bid-ask spread* is the difference between the current best buyer's price (ask value) and the best seller's price (bid value) of a given exchange. In less liquid markets, specifically in the most "exotic" cryptocurrency pairs (whose financial worth is typically in the USD cents), the percentual difference between the best buyer and the best seller may become large enough so that the common assumption of utilizing the closing value as both bid and ask values becomes quite far-fetched leading to a disparity with reality that may no longer become acceptable.

However, in the cryptocurrency exchange market, there is no single source providing free (or even for a reasonable monetary amount) bid-ask spread historical data. To the author's knowledge, enquiring specific exchanges or sources for the spread of the current instant (save for the Bitcoin/USD exchange market whose spread data is readily available, but becomes increasingly expensive as detail increases) is the only method of acquiring this data free of charge. Through this method, a database of instantaneous spread enquires could be built. Regardless, in order to assemble a decent amount of data, years of data gathering would be required. An alternative possibility would be utilizing a constant bid-ask spread for the whole time period, as done by Nakano et al. in [17]. Nonetheless, contrarily to Nakano et al.'s work where only the Bitcoin/USD exchange and its commonly available daily average bid-ask spread were analysed, in this work 100 different currency pairs are analysed with a precision of up to 1-min. By no means can an analogous procedure be applied to this work, specially for the most uncommon "exotic" cryptocurrencies which are significantly dissimilar from the most capitalized cryptocurrencies where the bid-ask spread becomes relatively negligible.

In this work, due to the unavailability of bid-ask spread values, only transaction fees are considered. It is preferable to be prudent rather than to undertake erroneous assumptions that could considerably increase the disparity from reality damaging the final conclusions. Nevertheless, it should be pointed that overlooking the importance of bid-ask spreads is not ideal and before applying this system in a real scenario, a database containing the spread values should be built and taken into account to fully validate this strategy.

In order to reduce risk, transaction fees are taken into account when calculating the target formulation. A positive variation, or 1, corresponds to when a closing price of a given time period deduced by the respective fee, is larger than the closing price of the previous time period, also with the respective fee taken into consideration. A negative variation, or 0, corresponds to the opposite case, when a closing price of a given time period deduced by the respective fee, is smaller than the closing price of the previous time period, also with the respective fee considered. As previously mentioned, in the rare cases where the signal has a null variation, the outcome of the previous instant is duplicated onto to the current one.

On the grounds of taking fees into account, a larger price gap is demanded between consecutive instants in order to enter a long position. Owing to this increased gap, small variations that could induce in losses are classified as negative variations rather than positive variations. Furthermore, positive variations can be translated into the possibility of an immediate profit (only a possibility because if the long position is held for long enough a negative trend may take place and turn this initial profit into a loss).

It may be noted that in theory there is a possibility that successive small increasing variations, which would individually result in losses due to the existing considered fees, could collectively generate a large enough increase resulting in a profit. Due to this formulation, possible profits from this occurrence are being discarded, as no individual variation would be large enough to trigger a positive variation which in turn would start a long position to collect the profits. However, in the cryptocurrency market, because of its volatility, the losses caused by small single positive variations amid negative variations far outbalances the gains from successive small variations that collectively generate a profit. Moreover, the resampling method mentioned in Sect. 3.2 is capable of grouping the original data to contain "individual" variations of a scale much larger than trading fee sized variations, hence, this situation will seldom be verified.

A vector called *vector y* (represented in Eq. (3.2)) contains the target classification for each sample in the dataset. Each entry follows a binomial probability distribution, i.e., $y \in \{0, 1\}$, where 1 symbolizes a positive or bullish signal variation and 0 symbolizes a negative or bearish signal variation.

Putting the previous process mathematically, for a given currency pair, being $close_t$ the closing price for a given time period t, and $close_{t-1}$ the closing price for the previous time period, $t - 1$, the target for the given time period, y_t, is defined as follows:

$$y_t = \begin{cases} 0, & \text{if } Close_{t-1} \times (1 + TradingFee) < Close_t \times (1 - TradingFee) \\ 1, & \text{if } Close_{t-1} \times (1 + TradingFee) > Close_t \times (1 - TradingFee) , \\ y_{t-1}, & \text{if } Close_{t-1} \times (1 + TradingFee) = Close_t \times (1 - TradingFee) \end{cases}$$

$$(3.2)$$

The only exception to the previous equation would be if the first two samples of the dataset fell under the last case, i.e., first close is equal to the second close with trading fees taken into account. As there is no previous entry to be duplicated into the actual one, to avoid risks a fixed negative classification, '0', is attributed instead.

In the end, all data samples have a corresponding entry in vector y except for the last sample as there is no posterior sample to evaluate whether the variation is positive or negative.

Having vector y defined, the problem can be written in a probabilistic way to employ the four learning algorithms: being x the observable features (technical indicators) at the current time and y_{t+1} the direction of the signal in the future, the probability for a given future direction knowing the current features, $P(y_{t+1}|x)$, for each learning algorithm may be calculated as explained in Sect. 2.2.2.

It's worth noting that only long positions are employed in this proposed system. The reason for this is because the utilized data was retrieved from Binance, and in accordance with this cryptocurrency exchange, short positions are not accepted (as of now). Therefore, to be truly faithful to a real case scenario, only long positions can be adopted by this system.

3.4.3 Classifiers Fitting and Predicting

This is the last component of the Machine Learning module. The objective of this component is, out of a dataset received as input, creating a model with the learned data by training each of the four classification algorithms and, in the end, creating a classifier ensemble with the output of each algorithm by averaging each sample. The four individual learning algorithms can be fit in whatever order.

Before the actual training procedure begins, both the features (matrix X) and the target vector (vector y), must be split into train dataset, used to train or fit the model, and test dataset, used to give an unbiased estimate of the generalization skill of the final tuned model. All training procedures, including cross-validation and grid searching, occur on the training datasets while the test datasets are kept isolated throughout the following procedures mentioned in this section. It is pivotal that each observation of the original complete dataset must belong to solely one set. Tainted testing datasets with data included as well in the training dataset, or vice-versa, manifest leaking of future data: training a given model with data also included in the testing dataset contradicts the concept of forecasting the future.

When developing a model for forecasting, it is obviously desirable to train and test the model on the longest possible datasets to cover the largest possible array of events. Therefore, in this proposed system both matrix X and vector y undergo multiple train-test splits, with the purpose of training and evaluating multiple models. Utilizing multiple test-train split sets has the benefit of providing a much longer testing interval which, despite being discontinuous, is in accordance with the prevention of future data leaking.

With the employed train-test split, each training set consists only of observations that occurred prior to the observations that form the test set. Thus, no future data leaking occurs in the construction of the forecast. For each currency pair in this system, 4 splits divide the time series in 5 equal sized intervals, i.e., each interval contains the same number of samples. It's worth noting that because the splits divide the number of samples, only the time rearranged series will be split in 5 time intervals with same time duration. The remaining resampling strategies, although having the same amount of samples in each interval, most likely will have clearly different durations for each of the 5 intervals. The test interval size is consistent for all iterations. This way, the calculated performance statistics of each tested model are consistent with each other and can be combined or compared. On the other hand, the train set successively increases in size with every iteration, i.e., it may be regarded as a superset of the train sets from previous iterations. Naturally, the first iteration is the exception, the training dataset consists only of the first interval. The test set corresponds to the immediate succeeding data interval relatively to the training dataset. For this reason, the generated classification models are less likely to become outdated in face of new distant testing data due to utilizing a single fixed classification model for the whole test dataset. Ultimately, when using this method, no future data leaking occurs and forecasting since a much earlier time point is enabled. This series of train-test splits are illustrated in Fig. 3.4.

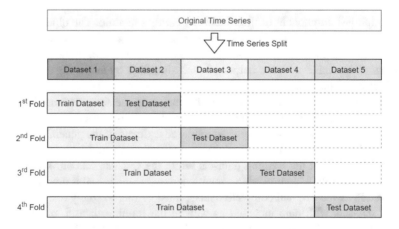

Fig. 3.4 Train-test split scheme

For each currency pair, because of the 4 different resampling methods utilized (explained in Sect. 3.2), the model training procedure has to be executed 4 different times. Additionally, 4 training intervals (as the fifth interval from the train-test split is only used for testing) are defined for each currency pair and this module is responsible for training one model for each of the 4 learning algorithms utilized. Hence, in total each currency pair has 64 models in total fitted. As soon as the boundaries between the split test and training sets are delimited, each classifier begins the model fitting procedure for each of the training intervals, one at a time. Throughout the training or fitting process, all classification algorithms share two aspects: the performance metric and the cross-validation method.

The first aspect to be defined is the used performance metric for model training. This evaluation metric defines which instance of the model is the best performing for each pair of the training and testing datasets. In this proposed system, all classification algorithms use negative log-loss or negative cross-entropy loss. For a binary classification case, the utilized formulae of this performance metric is represented in Eq. (3.3).

The reasoning behind this choice is that investment strategies profit from predicting the right label with high confidence. Accuracy accounts equally for an erroneous buy prediction with high probability and for an erroneous buy prediction with low probability [13]. For this reason, accuracy does not provide a realistic scoring of the classifier's performance. Conversely, log loss computes the log-likelihood of the classifier given the true class label, which takes the probabilities for each prediction into account, as can be seen in the following equation:

$$- log[P(y_t|y_p)] = -[y_t \times log(y_p) + (1 - y_t) \times log(1 - y_p)], \qquad (3.3)$$

where y_t corresponds to the true label, $\in \{0, 1\}$, and y_p corresponds to the estimated probability of $y_t = 1$.

As mentioned in Sect. 2.2.2 each learning algorithm requires the initialization of a set of parameters before the actual learning process begins. These parameters are called hyper-parameters [18]. In order to do so, the second aspect, cross-validation was introduced in this work's system. Cross-validation, is mainly used to determine the generalization error, i.e., its performance on unseen data of the model in order to prevent overfitting. Instead of partitioning the available data into three independent sets (train, validation and test), reducing the available number of samples to train and test the model, through usage of cross validation, multiple validation sets are used and later reused as training data. Take notice that any given data point once included in the training dataset remains in this same dataset for the remainder of the whole fitting procedure. One last advantage of cross-validation is that it makes full use of all available information both for training and testing, thus yielding more robust error estimates [19].

The type of cross-validation proposed for this system is commonly referred to as *time series cross-validation*. It is a variation of the commonly used *k-fold* cross-validation destined specifically for time series data samples. In the kth split, the first k folds are returned as train set and the $(k + 1)$th fold as validation set. Averaging the k obtained error measures yields an overall error measure that typically will be more robust than single measures [19]. This employed cross-validation is identical to the previously explained process of splitting train and test data, the only differences are that rather than only 4 splits, 10 splits are done and instead of splitting train and test data, train and validation data is being split. This is a simple notation difference. Although this approach may be computationally expensive, a more accurate performance estimate can be derived. On top of this, a strict and exclusive validation set is now dispensable, thus more data can be used on the training and/or testing intervals. Additionally, during cross-validation tuning of hyper-parameters can be carried out. In this work, a simple grid search was carried out in this proposed system.

Grid-searching is a simple exhaustive search through a manually specified subspace of values with the purpose of finding the best values for each hyper-parameter. During the model training step, in each fold, various instances of each learning algorithm are trained with all the possible hyper-parameter combinations and tested on the respective validation set. This way, the performance according to the negative log-loss metric is obtained on unseen data for each hyper-parameter combination.

The results for each hyper-parameter combination are averaged through all validation sets. The instance of the model containing the average highest cross-validated scoring parameters is then refit on the whole training dataset to measure its performance on the whole training dataset continuously. This best instance is also later used as the final model to issue the final predictions on the test dataset.

For each classifier in the learning procedure, the value for each parameter was chosen with the objective of maximizing the predictive performance. To accomplish this, multiple combinations of parameters were grid searched for each classifier in order to find the best parameters. While some parameters are still grid searched in this final version of the system, in RF and SVM's case none are. These two are the slowest algorithms, hence, because both computational power and time were limited for the creation of this study, only the best values obtained for each parameter are directly applied. In spite of being computationally expensive, the grid search method would not take much longer than advanced methods because in this system not too many parameters are grid searched for each of the used methodologies.

To conclude the training process, the last aspect are the additional specific parameters of each individual classifier used. A description of the most relevant specific parameters and aspects of each classification algorithm follow:

- **Logistic Regression**: First of all, even though it is uncommon for the datasets utilized in this thesis to be remarkably imbalanced, with the utilized data it is common to witness a 60%-40% level of imbalance between the two class labels. Hence, instead of using a unitary weight for both classes, the weights for each class are adjusted inversely proportional to class frequencies in order to put more emphasis on the most infrequent classes. In other words, each mistake is penalized differently for each class as follows:

$$w_j = \frac{n}{k \times n_j}, \tag{3.4}$$

where w_j is the weight to class j, n is the total number of samples, n_j is the number of observations of class j and k is the total number of classes. This way, the most uncommon class is penalized the most in order to improve class performance of the minority.

In this proposed system L2 penalization is used and the C parameter is grid searched, both parameters were discussed in Sect. 2.2.2. Finally, multiple solvers for the optimization were experimented with. However, the difference between results was negligible, the "saga" algorithm [20], a variant of the Stochastic Average Gradient descent algorithm, was used due to its faster convergence and consequently shorter execution time.

- **Random Forest**: Identically to LR proportional weights on errors are added to control class imbalance. In terms of time intensity, this classifier is a close second, only falling behind SVM, hence grid searching was not utilized in the end version. The final version of this model contained 400 trees in the forest and both the minimum number of samples required to be a leaf node and minimum number of samples required to split an internal node were chosen to be 9. The splitting criteria utilized is the Gini index mentioned in Sect. 2.2.2 and the minimum impurity threshold for a split is 10^{-7}. As mentioned in the same section, each split instead of looking into all features, only considers a specific set. The amount of features considered correspond to the formula used by default, the square root of all available features [6].

- **Gradient Decision Tree Boosting**: Once again, proportional weights on errors are added to control class imbalance identically to LR. For this learning method 100 boosted trees were fit with a maximum depth of 3. The minimum loss reduction required to make a partition on a node of the tree, the minimum sum of instance weight needed in a child and the L2 regularization strength were grid searched. The minimum loss reduction required to make a further partition on a leaf node of the tree was chosen as 1. Finally, the step size shrinkage used was defined as 0.01.

- **Support Vector Machine**: Similarly to the previous methods, proportional weights on errors are added to control class imbalance. The C parameter explained in Sect. 2.2.2 was defined as 1. As this method additionally employs Platt scaling, which takes a considerable amount of time, a linear kernel was used principally to avoid an even larger computational complexity and consequently slower execution. For this same reason, the maximum number of iterations had to be limited to 10000 in this classifier and still, it was by far the most time intensive algorithm. It had to be limited as the computational power available for this thesis too, was limited. Hence, the search for the optimal hyper-parameters is somewhat restrained.

The training process of the four binary classifiers is the most time intensive procedure of the whole system. Therefore, before all 4 classification algorithms are trained, it is firstly verified if a previously fit model for the specific cryptocurrency pair and train interval (by verifying the starting and ending dates to a precision of minutes) has been saved into memory. If so, there is no need to retrain that model, the existent one is simply loaded up. Otherwise, a model is fit for the train interval and saved to be later used. Additionally, the weight for each feature is written in a file for statistical purposes.

3.4.4   Forecasting and Ensemble Voting

At this point, having a model for a specific training dataset either fit or loaded into memory, may the second part of supervised learning, the forecasting, commence. For each specific training dataset, all learning algorithms apply the fit model on the respective testing dataset. A probability estimate of the respective class labels for each sample of the test dataset is generated with each algorithm and stored to be later used on the weighted averaging of the 4 classification models. Only one of these odds needs to be saved as these two events are mutually exclusive and the classification is binary and collectively exhaustive, hence their sum always equals 100%. In this work, each obtained series containing the probability of each sample in a given test dataset being classified as 1 is named the *trading signal*.

The classification odds for each test sample of the 4 models, were combined through *soft majority voting* into an ensemble. The driving principle behind majority voting is to build several estimators independently and then to average their predictions. In soft majority voting, the probabilities for the prediction of each class label of the test dataset previously calculated are averaged. This way, soft majority voting takes into account how certain each classifier is, rather than just a binary input from each classifier as in *hard majority voting*.

At last, this module outputs the five different trading signals generated containing the predicted data. Four trading signals are originated from the individual learning algorithms while the last is originated from the unweighted average of these four. All the trading signals have the same time frame, and will be simulated in a real world environment in the next module, the investment simulator.

3.5   Investment Simulator Module

The investment simulator model is responsible for translating the five trading signals obtained from predicted data in the Machine Learning module received as input, into market orders with the purpose of simulating the forecasts' performance in a real market environment.

The trading signal received as input, consists on a series containing the probability of each sample being classified as 1, according to the target formulation specified in Sect. 3.4.2. The trading signal is firstly converted into a class label series according to the largest likelihood. In this system the trading signal received as input contains the probability of a given sample being classified with the label *1*, hence, a probability of 50% or over is classified as *1*. Conversely, if the odds are under 50%, then a class label *0* is assigned.

After the trading signal has been converted into the class label series, each entry of this series is interpreted as follows:

- A class label of *1* meaning the currency is forecast to be bullish in the next instant represents a *buy* or *hold* signal. If the system is currently out of the market, i.e., not holding units of base currency, the system opens a long position at the closing price of the corresponding trading period. All available quote currency units are applied by the system when opening a new long position. On the other hand, if the system is currently with a long position active, meaning it is only holding units of base currency, this class label orders the system not to sell, keeping all its units of base currency.
- A class label of *0* meaning the currency is forecast to be bearish in the next instant represents a *sell* or *out* signal. If the system is currently not holding units of base currency, this signal orders the system not to buy and stay out of the market, keeping the same amount of capital. On the other hand, if the system is currently holding units of base currency, an order is issued by the system to sell all its units of base currency, i.e., convert them into quote currency.

To simulate the market orders, backtest trading is employed in this work. Backtest trading is the process of applying a trading strategy to historical data in order to measure how accurately the strategy would have predicted historical results and performed in an actual market situation, without risking actual capital. The assumption is that if the strategy has worked with past data, there is a good chance of it working again in the future and conversely, if the strategy has not worked well in the past, it probably will not work well in the future. Because this backtest process utilizes historical data, the following two assumptions must be imposed for this proposed system:

1. Market liquidity: The markets have enough liquidity to conclude each trade placed by the system instantaneously and at the current price of their placement.
2. Capital impact: The capital invested by the algorithm has no influence on the market as it is relatively insignificant.

This module starts with a fixed capital for investment (by default it's one unit of quote currency) and invests that capital according to the previously mentioned interpretation of the class label series. Each of the five trading signals outputted from the previous module are independently converted and are simulated continuously throughout the entire testing period. It's worth reminding that trading positions are contracted with closing prices to represent the worst case scenario. In other words, trades are only executed at the end of each candle in order to enact the waited time experienced in reality.

The fees introduced in Sect. 3.4.2 are taken into consideration in the trading simulator as well with the purpose of reproducing a real market. Each trading order (both buy and sell) made by the simulator takes into account Binance's general trading fee of 0.1% [16].

The behaviour of this module, with fees included, is exemplified in Fig. 3.5. In this figure the red columns represent period intervals where the investment strategy

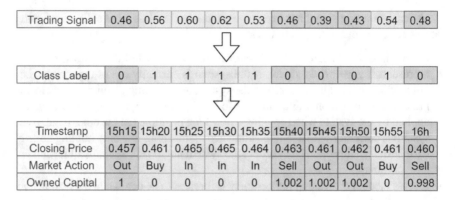

| Trading Signal | 0.46 | 0.56 | 0.60 | 0.62 | 0.53 | 0.46 | 0.39 | 0.43 | 0.54 | 0.48 |

| Class Label | 0 | 1 | 1 | 1 | 1 | 0 | 0 | 0 | 1 | 0 |

Timestamp	15h15	15h20	15h25	15h30	15h35	15h40	15h45	15h50	15h55	16h
Closing Price	0.457	0.461	0.465	0.465	0.464	0.463	0.461	0.462	0.461	0.460
Market Action	Out	Buy	In	In	In	Sell	Out	Out	Buy	Sell
Owned Capital	1	0	0	0	0	1.002	1.002	1.002	0	0.998

Fig. 3.5 Execution example of the trading module

is outside or should immediately leave the market and, conversely, the green columns represent inside or should enter the market immediately. The represented *Trading Signal* row contains the probability of each sample being classified as 1. This series of probabilities, as was previously explained, is easily converted into a series containing class labels, which is represented in the second row. The class label series dictates what should be done in the market under analysis, briefly summarized in the last four rows of the figure. Note that the *Owned Capital* row corresponds to the investment strategy's owned capital in quote currency units at each given instant.

Lastly, as a method to control risk on a trade-by-trade basis, in this system stop-loss orders were implemented. This method of approaching risk leads to traders placing orders to sell/buy securities to cover open long/short positions when losses cross predetermined thresholds. An initial stop-loss is ordered whenever a market entry, or buy, is placed. If the price of the security falls surpassing a specified threshold, then a stop-loss is triggered and the position is sold at a loss. If the stop price is further away from the current price, then there is potentially more money at risk, however, there is a lower chance that the stop will be triggered by noise. On the other hand, the use of tight stops may be degrading towards long term performance [21] as well. Clearly, the closer the stop is to the actual price of the security, the less money will be lost if the price falls, but the more likely it is that a smaller price fluctuation or random noise will trigger the stop. Hence, the stop-loss percentage must be chosen so that only in drastic cases is it triggered. For this system a default stop-loss of 20% was chosen.

In the end, a series of metrics, revealed and explained in Sect. 4.2, are calculated in this simulator and written into a file. Finally, each prediction and its respective metrics can be plotted and stored into memory.

3.6 Conclusions

This chapter was intended to give an insight of the proposed system's architecture. The main architectural decisions, many of them based on the literature reviewed in Chap. 2, are reasoned and further described throughout this chapter.

To sum up, this system is composed of four main models: the Data, Technical, Machine Learning and Investor Simulator modules; each of these modules is interconnected to the remaining ones and is indispensable towards the development of the final trading signal. These main modules are described in detail throughout this chapter. Notice that several aspects of these system are easily modifiable or extendible, namely the utilized technical indicators and their parameters, the threshold that defines the resampling frequency, the parameters that characterize each individual learning algorithm, the stop-loss activation parameter, etc. In Chap. 4, the system explained throughout this chapter is evaluated in full detail.

References

1. Wolpert DH, Macready WG (1997) No free lunch theorems for optimization. IEEE Trans Evol Comput 1(1):67–82
2. Zhou ZH (2012) Ensemble methods: foundations and algorithms. CRC Press, Boca Raton
3. Raschka S (2015) Python machine learning. Packt Publishing Ltd., Birmingham
4. Caruana R, Niculescu-Mizil A (2006) An empirical comparison of supervised learning algorithms. In: Proceedings of the 23rd international conference on machine learning, pp 161–168
5. Nielsen D (2016) Tree boosting with XGBoost-why does XGBoost win "every" machine learning competition? Master's thesis, NTNU
6. Hastie T, Tibshirani R, Friedman J (2009) The elements of statistical learning: data mining, inference, and prediction. Springer Science & Business Media, New York
7. Cavalcante RC, Brasileiro RC, Souza VL, Nobrega JP, Oliveira AL (2016) Computational intelligence and financial markets: a survey and future directions. Expert Syst Appl 55:194–211
8. Lütkepohl H, Xu F (2012) The role of the log transformation in forecasting economic variables. Empir Econ 42(3):619–638
9. Sharma JL, Kennedy RE (1977) A comparative analysis of stock price behavior on the Bombay, London, and New York stock exchanges. J Financ Quant Anal 391–413
10. Arratia A (2014) Statistics of financial time series. Computational finance. Atlantis Press, Paris, pp 37–70
11. Fama EF (1965) The behavior of stock-market prices. J Bus 38(1):34–105
12. Tsay RS (2005) Analysis of financial time series, vol 543. Wiley, New York
13. De Prado ML (2018) Advances in financial machine learning. Wiley, New York
14. Teixeira LA, De Oliveira ALI (2010) A method for automatic stock trading combining technical analysis and nearest neighbor classification. Expert Syst Appl 37(10):6885–6890
15. Yang C, Odvody GN, Fernandez CJ, Landivar JA, Minzenmayer RR, Nichols RL (2015) Evaluating unsupervised and supervised image classification methods for mapping cotton root rot. Precis Agric 16(2):201–215
16. Binance. Binance fee schedule. https://www.binance.com/en/fee/schedule. Accessed 02 Sept 2020
17. Nakano M, Takahashi A, Takahashi S (2018) Bitcoin technical trading with artificial neural network. Phys A: Stat Mech Appl 510:587–609

18. Murphy KP (2012) Machine learning: a probabilistic perspective. MIT Press, Cambridge
19. Bergmeir C, Benítez JM (2012) On the use of cross-validation for time series predictor evaluation. Inf Sci 191:192–213
20. Defazio A, Bach F, Lacoste-Julien S (2014) SAGA: a fast incremental gradient method with support for non-strongly convex composite objectives. In: Advances in neural information processing systems, pp 1646–1654
21. Chande TS (2001) Beyond technical analysis: how to develop and implement a winning trading system, vol 101. Wiley, New York

Chapter 4
Results

4.1 Financial Data

The used data was extracted from Binance's API owing to its accessibility and to the extensive list of cryptocurrency pairs nowhere else offered. Particularly pairs containing non-mainstream cryptocurrencies whose exceptional volatility could be exploited by this work's system. Nevertheless, instead of using the nearly 400 currently available pairs in Binance, only the 100 pairs with the most traded volume were selected. This selection filters out many pairs that have been recently listed and don't have a large enough dataset. When utilizing small data sets, among other several undesired effects, outliers and noise have a larger impact to the point of becoming troublesome and increasing the risk of overfitting a model. Another reason for selecting the pairs with most volume is that in a real investment scenario, the market orders generated by this algorithm are more likely to be carried out immediately and to be insignificant relatively to the remaining orders. Accordingly, it becomes more unlikely to saturate the market or to even have a noticeable impact. Thus, through this filtration, it is more likely that the selected markets are in conformity with the backtest trading assumptions expressed in Sect. 3.5.

For the sake of finding the most actively traded pairs, the total volumes must be compared with each other in the same currency unit. Due to the large availability of data, the US Dollar was chosen as the comparison currency for this purpose. In order to convert and compare the volumes, firstly, for each pair, the volume of each data point was converted into USD according to the respective price at each data point's time instant. Secondly, the sum of the volume in USD for all data points of each pair was calculated, outputting the total volume in USD. Finally, the list of pairs was ordered by total volume in USD.

It should be noted that in the markets MTHBTC, MTLBTC and SNMBTC, who all placed in the selection of 100 markets, the returns obtained were in the order of hundred thousand percent. The bid-ask difference in these markets is evidently too large to be overlooked. Hence these three markets were simply swapped for the next

© The Author(s), under exclusive license to Springer Nature Switzerland AG 2021
T. A. Borges and R. Neves, *Financial Data Resampling for Machine Learning Based Trading*, SpringerBriefs in Computational Intelligence, https://doi.org/10.1007/978-3-030-68379-5_4

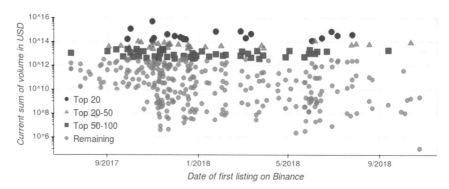

Fig. 4.1 Current total volume in USD as a function of the respective introduction date in Binance for each cryptocurrency pair

three markets with most volume. This was done to avoid tampering the final and average results. Figure 4.1 contains the introduction date and current volume in USD of the 100 markets used. Three additional markets will also be swapped, as will be explained in Sect. 4.4.1, this figure also takes into account this additional swap.

In order to use the maximum amount of available data, no common starting date was chosen for all currency pairs. Each currency pair's data begins at the moment Binance started trading and recording the specific currency pair. The ending date, on the other hand, is fixed at 30th October 2018 at 00h00. Out of the used cryptocurrency pairs, the largest pairs originally contain 676827 trading periods while the smallest pairs contain 68671 trading periods. Each trading period, as previously mentioned, has the duration of 1 min. The starting date varies from 14th July 2017 (Binance's official launch date) for the oldest pairs, up to 12th October 2018 for the most recently introduced pair. Surprisingly, pairs containing two well known currencies such as ETH/BTC (Ethereum/Bitcoin) or LTC/BTC (Litecoin/Bitcoin) did not enter the top 100 cryptocurrency pairs with the most volume. This is possibly because these well known pairs have been actively traded for longer than Binance's creation, hence users may have preferred not to change their marketplace. Moreover, one main objective of Binance, contrarily to many other exchanges, has always been to offer a large variety of cryptocurrencies, many of them are smaller and considered as "exotic". Binance's extensive variety of rarely offered cryptocurrencies is an attractive and differentiating factor for investors when picking an exchange.

The original data for the 100 selected currency pairs, prior to its usage in forecasting, must be resampled. As was explained in Sect. 3.2, to carry out the multiple resampling procedures, a fixed percentual variation threshold must be picked to define the approximate final size of each candle in the final dataset. In the subsequent results, a fixed percentage of 10% was chosen for the resampling procedures according to the following reasons:

- Trading Fees: This value ought to be greater than the constant trading fees in order to avoid loss of profits as was described in Sect. 3.4.2. Hence, the minimum considered percentage must be larger than 0.1%.
- Volatility: As mentioned in Sect. 2.1 the cryptocurrency exchange market is quite volatile relatively to the remaining financial market world. Therefore, resampling the original data to variations in the order of 10% still yielded positive results and not too much information was lost. Moreover this value is much larger than the trading fee which agrees with the previous point.
- Computational Power: Due to limited computational power, even with this seemingly high percentage of 10%, running the whole system for the 100 selected cryptocurrency pairs took several days.

4.2 Evaluation Metrics

As was mentioned before, the main goal of this proposed system is maximizing the negative logarithmic loss and returns while minimizing the associated risk of this work's trading strategy. With this goal in mind, the following metrics of market performance were additionally calculated for each currency pair with the intention of providing a better analysis of the obtained results:

4.2.1 Return on Investment

The return on investment (ROI) measures the amount of return gained or lost in an investment relative to the initially invested amount. This metric is expressed as a percentage and, thus, can be used to compare the profitability between different investments or strategies during a given period of time. The simple standard formula of this metric is represented as follows:

$$ROI = \frac{FinalCapital - InitialCapital}{InitialCapital} \times 100\%, \qquad (4.1)$$

where $FinalCapital$ corresponds to the capital obtained from the investment bought with $InitialCapital$.

4.2.2 Maximum Drawdown (MDD)

The maximum drawdown measures the maximum decline, from a peak to a through before a new peak is attained, of a given investment. This indicator is used to assess the relative downside risk of investments or strategies with focus on capital preservation during a given time period [1]. This metric is calculated as follows:

$$MDD = \max_{t \in (StartDate, EndDate)} [\max_{t \in (StartDate, T)} (ROI_t) - ROI_T], \qquad (4.2)$$

where ROI corresponds to the return on investment at the subscript's point in time and $\max_{t \in (StartDate, T)} (ROI_t)$ corresponds to the highest peak from the starting point until the instant T. As a whole, Eq. (4.2) can be seen as the maximum of all possible drawdowns. This equation may posteriorly be divided by the peak value correspondent to the final MDD to convert the value into a percentage. In this work, as is customary, MDD it is quoted as a percentage of the peak value.

One final remark is that, in this work cryptocurrency markets are being analysed. Owing to their inherent volatility, this metric most often will be higher relatively to other systems who handle more stable financial markets.

4.2.3 Sharpe Ratio

The Sharpe Ratio is a method for calculating the risk-adjusted return. This ratio describes the excess return received for holding a given investment with a specific risk, conveying if the investment's return is due to a smart decision or if it is the result of a higher level of risk. The sharpe ratio is calculated as follows:

$$SharpeRatio = \frac{ROI - R_f}{\sigma}, \qquad (4.3)$$

where ROI corresponds to the return on investment, R_f is the current risk-free rate, 3.5%, and σ is the standard deviation of the investment's excess return. Generally, the greater the value of the Sharpe ratio, the more attractive the risk-adjusted return is. Finally, it's worth noting that while for most financial markets (stocks, bonds, funds, etc.) 252 days in a year are used, for the cryptocurrency exchange market, 365 in a year are considered because these are continuously available for trading at all times.

4.2.4 Sortino Ratio

The Sortino Ratio is a modification of the Sharpe ratio metric also used for measuring the risk-adjusted return of a strategy. In contrast to Sharpe ratio where all returns variations are accounted, the Sortino ratio includes only negative variations. Upside variations are beneficial to investors and are not a factor to worry about. Due to this fact, when accounting for risk, this metric takes out of the equation these positive variations. In this sense, the Sortino ratio has the objective of presenting a more realistic risk measure. The formula for calculating this metric is identical to the formula displayed in Eq. (4.3), however, the denominator corresponds only to the standard deviation values observed during periods of negative performance, also known as downside deviation.

4.2.5 Additional Performance Parameters

Besides the four presented metrics used to evaluate the performance of an investor, the following parameters are also used in the classification of this proposed system:

- Percentage of periods in market: Percentage of time periods where a long position was in effect out of all the available time periods of the testing set. A percentage is used because the analysed markets in this work do not contain the same number of test periods. Hence, a direct comparison between periods in market would be meaningless, a relative comparison had to be resorted to. This metric is intended to measure how active this work's proposed system is in a market. Due to the characteristically high volatility of the cryptocurrency markets, it is preferable that a strategy contains a low percentage of periods in market. The investor becomes less susceptible to risk;
- Percentage of profitable positions: This parameter is calculated by dividing the number of trades that generated a profit (with fees included), by the total number of trades. It is worth noting that this probability is complementary to the percentage of non-profitable positions (positions that yielded a precisely null variation in returns or losses), a similarly relevant measure;
- Average Profit per position: Average percentual profit or loss per position, considering all trades within a specific time period. To calculate this parameter, the net percentual profit is divided by the total number of positions assumed;
- Largest Percentual Gain: Most profitable position taken in the period of analysis, calculated as a percentage. It is the largest obtained result of $(PriceSold - PurchasePrice)/PurchasePrice$;
- Largest Percentual Loss: Similarly to the greatest profit, it is also relevant to determine the greatest loss. It is the lowest obtained result of the equation mentioned in the previous parameter.

4.3 Additional Buy and Hold Strategy

In order to validate this system, besides testing with real market data through backtest trading and analysing the results with the just introduced metrics, a previously mentioned investment strategy, the Buy and Hold (B&H) is applied. According to the Efficient Markets theory addressed in Chap. 1, prices are independent of each other, hence no profit can be made from information based trading. In conformity with this theory, the best strategy is employing a buy and hold strategy, regardless of market fluctuations. Due to the limited number of existent solutions for trading in multiple cryptocurrency pair markets and to put the Efficient markets theory to the test, B&H was defined as a benchmark strategy intended to be an additional term of comparison for this work's proposed system.

With the purpose of simulating this strategy, this work's system had to somehow be able of interpreting this strategy. In order to do so, this strategy was converted into a trading signal following the same rules of the previous five used trading signals. Therefore, the B&H trading signal consists on a vector of 1's, except for the last entry which is a 0 (the trading signal coincides with the class label series). Thus, this trading signal establishes that the system enters a long position on the first period of the first test dataset. This instant purposefully coincides with the first period where the remaining trading signals, derived from the machine learning module, are able of entering the market, this way all signals fairly have the same winning potential. The system is kept inside the market until the very last period where the starting long position is closed. It is assumed that this strategy assumes with a certainty of 100% that each sample should be classified as 1, or 0 for the last sample's case.

It's worth noting that the accuracy and negative log-loss metrics are included for this strategy in order to provide a perspective on how the market itself reacts to the target formulation developed in this work's system. The accuracy metric simply indicates how many entries of the financial time series were assigned a label 1, while the NLL metric indicates the uncertainty of the prediction based on how much it varies from the actual label.

4.4   Case Studies

In this section, the following case studies are described to evaluate the developed system:

- Case Study A: Prior to analysing in detail the actual results of this work's proposed system, a few cases of illiquid markets worth noting are identified and explained.
- Case Study B: Comparison between B&H strategy, each of the four different machine learning algorithms and the ensemble voting classifier.
- Case Study C: Analysis of time resampling.
- Case Study D: Analysis of logarithmic amount resampling.
- Case Study E: Analysis of percentage resampling.
- Case Study F: Analysis of amount resampling.

The case studies C, D, E and F are ordered according to EV's average final ROI, because as will be seen in case study B, this method originates the trading signals with best overall performance. Notwithstanding the existence of a fixed execution order in the system where time resampling has to be the last method to be executed, the final results obtained from the different resampling methods are independent. Additionally, time resampling is undoubtedly the most widely used method of resampling,

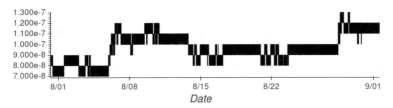

Fig. 4.2 OHLC representation of the market values during the month of August 2018 for the currency pair HOTBTC (Holo/Bitcoin) with 1-min precision

therefore, it ought to be the first sampling method analysed to be considered as the comparison baseline towards the other three case studies.

4.4.1 Case Study A—Illiquid Markets

Before going into detail about the actual results of this work's system, it is worth noting these specific occurrences. A few cryptocurrency pairs from the 100 selected, provided suspiciously high ROIs. Namely, the currency pairs HOTBTC, NPXSBTC and DENTBTC, who all placed within the top 10 pairs with most volume, obtained ROI values in the orders of over a million million percent. In reality this ROI is obviously unobtainable, hence these specific pairs were further investigated.

These pairs have open and close values fixed for several hours at a time and on top of that the percentual difference between open and close values is in the order of 5% or above. These markets have the potential of generating huge profits. Predicting the price direction of the next instant is not a difficult task: the algorithm simply goes long on the relative or local low value and closes the position as soon as a higher value is reached. Because the open and close values are locked for long periods of time, rarely is the algorithm forced to close the position at an equal (losing profits due to fees) or lower value relatively to the value in which the long position was initiated. This work's proposed system enters and closes a long position once every 5–30 min obtaining profits in the orders of 5% the majority of times. Keeping this strategy for a few months would yield monumental profits.

The Holo-Bitcoin (HOTBTC) market illustrated in Fig. 4.2 exemplifies this situation. Holo's value is around the minimum divisible amount of Bitcoins,[1] thus the smallest possible variations for the open-close values are over 5% in size. Note that the represented candles are all dark-coloured because too many small candles are being displayed. The used software is incapable of rendering colour at such great detail.

After a thorough analysis it was concluded that in theory this occurrence seems to be correct, however, in practice it would be impossible to achieve. In this work's system, due to the backtest process employed, it was assumed that the invested markets

---

[1]The smallest divisible unit of Bitcoin is one Satoshi unit corresponding to 10^{-8} of a Bitcoin.

have the liquidity to accept instantaneously buy and sell orders given by this system. Instantaneous acceptance of an order is more probable in markets with large traded volumes, which is why such markets were preferred to test this system on. However, in these 3 cryptocurrency pairs the markets are just not liquid enough. According to Binance's cryptocurrency exchange, any buy or sell transaction would take on average a few days, whereas this work's system on average intends to enter and close a long position within 5–30 min. Due to the impossibility of achieving these buy and sell rhythms, in reality reaching the obtained ROI values would be unfeasible, consequently, these three cryptocurrency pairs were excluded. Nonetheless, for consistency's sake, to keep the number of analysed markets an even number, these three markets were swapped for the three next markets with most volume.

As a side note, it seems that these 3 (and possibly more) cryptocurrency pairs may potentially be caught in a loop of slow but easy and guaranteed profit. It could be for this reason that all 3 pairs have such a large volume in Binance. In any case, this falls on the realm of speculation and is way beyond the scope of this work.

4.4.2 Case Study B—Methodologies Overview

First of all, it's worth noting this case study is intended to provide a general idea of the overall performance obtained with each different methodology independently of the resampling method utilized. Contrarily to the succeeding case studies where in depth results are presented, this current case study had to sacrifice detail in order to present a broader view over the system's results. To achieve this, whilst utilizing the buy and hold strategy as benchmark, the five trading signals generated by the four individual learning algorithms and the ensemble voting method are individually averaged and subsequently compared against each other.

The shown results were obtained by executing this work's proposed system exactly as explained in Chap. 3. Each of the five methodologies built 4 different trading signals for each cryptocurrency pair, one for each resampling method. A total of 100 cryptocurrency pairs from Binance are analysed, as mentioned in Sect. 4.1, hence a total of 2400 (100 Pairs×4 resampling Methods×6 strategies (5 Classifiers+B&H)) individual trading signals are analysed. A general overview of the results is presented in Figs. 4.3 and 4.4. Each of these figures contains five radial bar plots, one for each methodology, with a general overview of the ROI and accuracy metrics. In Fig. 4.3 each single blue bar represents the final ROI for a given market. If a blue bar is placed over a green background, the respective methodology acquired a larger ROI than B&H did for the given market, while a red background represents the opposite. It is only worth adding that the B&H strategy's ROI had no comparison method, hence the background is clear. Analogously, in Fig. 4.4 each single blue bar represents the average accuracy obtained during the testing period for a given market, a green background means that the accuracy obtained for the respective market is larger than 50% and a red background the opposite.

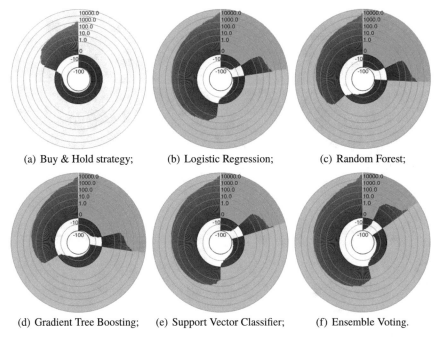

(a) Buy & Hold strategy;   (b) Logistic Regression;   (c) Random Forest;

(d) Gradient Tree Boosting;   (e) Support Vector Classifier;   (f) Ensemble Voting.

● ROI worse than B&H strategy;   ● ROI better than B&H strategy.

Fig. 4.3 Radial bar charts containing a general overview of the ROI's obtained, measured in percentage

Since the remaining metrics are not visually as informative or intuitive as the ROI or accuracy metrics, their average values are summarized in Table 4.1 instead, which is sufficient to demonstrate the general overview of the obtained results for each methodology. Once again, the B&H strategy is additionally represented as the baseline for comparison.

Through analysis of Figs. 4.3 and 4.4 and Table 4.1, the following observations can be made:

- **Buy and Hold**, the benchmark strategy, as initially expected, obtains the worse values for most metrics in the great majority of markets. This simple approach to investing obtains a significantly lower ROI and predictive power. Furthermore, on average this strategy obtains the worst values for the risk metrics (Sharpe ratio, Sortino ratio and MDD). After all, contrarily to the other strategies derived from learning algorithms, B&H, in any event, blindly stays inside the market until the last sample is reached.
- **Logistic Regression** was favourably unexpected. Out of the individual learning algorithms it placed as second best performer in terms of ROI and profit per position, only being narrowly exceeded by the SVC in these metrics. On top of that, out of all methodologies it obtained the best percentage of profitable positions. At

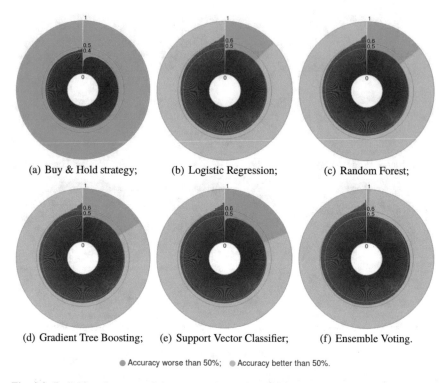

(a) Buy & Hold strategy; (b) Logistic Regression; (c) Random Forest;

(d) Gradient Tree Boosting; (e) Support Vector Classifier; (f) Ensemble Voting.

● Accuracy worse than 50%; ● Accuracy better than 50%.

Fig. 4.4 Radial bar charts containing a general overview of the accuracies obtained

the same time, this learning algorithm on average also obtained the worst negative log-loss. The fact that this algorithm on average is the individual learning algorithm most frequently with a long position active in market coupled with obtaining the worst negative log-loss, suggests that LR yields a risky trading signal.

- **Random Forest** yielded surprisingly underwhelming results. Despite being clearly superior to the benchmark strategy, the results did not exceed or even match the initial expectations whatsoever. The obtained values for this learning algorithm are sub-par. This algorithm obtained the worst ROI and the second worst scores for MDD and Sharpe and Sortino ratios out of the individual algorithms. All in all, this learning algorithm by itself generates a risky trading signal with the lowest returns relatively to the remaining learning algorithms.
- **Gradient Decision Tree Boosting** also had a sub-par performance. For most metrics the same conclusions of RF can be taken, the exceptions are the slightly larger ROI and negative log-loss. In fact, the trading signals generated from this algorithm on average have the highest negative log-loss out of all individual learning methods. On the other hand, the largest MDD as well as lowest Sharpe and Sortino ratios were also obtained. Apart from B&H, this method yields the riskiest trading signals.

Table 4.1 Comparison between the Buy & Hold strategy and each of the five methodologies employed

Parameter	B&H	LR	RF	GTB	SVC	Ensemble voting
Average obtained results (all markets and resampling methods are considered)						
Final ROI	−10.45%	518.5%	294.6%	335.4%	537.8%	**614.9%**
Accuracy	40.2%	53.50%	53.62%	53.51%	53.38%	**56.28%**
Negative log-loss	−20.6	−0.7031	−0.6992	−0.6918	−0.6975	**−0.6829**
Periods in market	100%	56.02%	52.91%	55.04%	50.39%	**39.87%**
Profitable positions	19.5%	**60.22%**	56.17%	58.30%	58.77%	57.61%
Profit per position	−10.46%	0.565%	0.307%	0.325%	0.610%	**0.686%**
Largest gain	**35.6%**	17.64%	17.71%	17.28%	18.39%	15.02%
Largest loss	−46.0%	−14.39%	−15.01%	−15.14%	−14.51%	**−13.15%**
Max drawdown	79.9%	57.6%	60.6%	62.0%	54.7%	**49.3%**
Annual sharpe ratio	−0.164	0.769	0.413	0.312	0.848	**0.945**
Annual sortino ratio	0.169	2.374	1.665	1.407	2.568	**2.821**

- **Support Vector Classifier** performed as expected, it is clearly one of the top tier learning algorithms employed in this system. Concerning riskiness, in addition to a relatively good Sharpe and Sortino ratios as well as MDD, the value obtained for percentage of periods in market is the lowest out of all individual learning algorithms. On top of this, the average largest gains for a method based on machine learning are obtained with this classifier and a ROI better than any of the previous classifiers is obtained on average with this method. All things considered, out of all individual classifiers, the SVC usually generates the trading signals with the less risk and highest ROI associated.
- **Ensemble Voting** is by far the most robust alternative. Employing this most computationally expensive and time intensive method consistently yields the best payoff. As can be seen, the trading signal obtained from EV obtains the best values for the majority of the metrics. This method obtains top scores for the metrics associated with forecasting performance: accuracy and negative log-loss, suggesting that this is the best predictor. The remaining metrics also suggest the superiority of this method, namely due to obtaining on average the largest ROI. Within a specific cryptocurrency pair, this methodology seldom obtains the worst ROI. In fact, through observation of the individual results, in a specific cryptocurrency pair the ROI obtained by EV is usually well above the average of the final obtained ROI

of the other four methods. Regarding risk, this method obtained top scores for the MDD, both Sharpe and Sortino ratio, the smallest percentage of periods in market, largest average profit per position and the smallest largest loss values. All in all, relatively to the remaining signals of this system, the trading signals generated by this method are by far the less risky and the most profitable. It is worth noting that the trading signals generated by this method did not obtain the top score for largest average gains most likely due to the more conservative stance as can be confirmed by the relatively small amount of periods in market.

The obtained accuracies are on par or, for the case of ensemble voting, exceed the accuracy of most papers regarding cryptocurrency exchange market forecasting throughout the state-of-the art (Sect. 2.3 and Table 2.1). The risk measures and returns on investment are also superior on average. Only Mallqui and Fernandes [2] obtained an absolute superior accuracy. It should be noted that in their work only one single market was analysed and as can be observed in Fig. 4.4, independently of the resampling method, this system is also capable of achieving such levels of accuracy.

To conclude, there is no clear performance hierarchy for each of the four individual learning algorithms. Nonetheless, the ensemble voting methodology is the top performer out of all five methods. In any case, all five methodologies clearly outperform the plain Buy and Hold strategy. This alone is not indisputable evidence to entirely disprove the efficient market theory (mentioned in Chap. 1), but is enough to suggest that the analysed cryptocurrency exchange markets possess characteristics that oppose the Efficient Market Hypothesis.

One final note is that managing a portfolio with only a subset of all these cryptocurrencies could be beneficial for the outcome of the overall system. With an effective portfolio management, markets that no matter the resampling or classification method employed, consistently yield negative results, could potentially be excluded. Thereby, only the subset of cryptocurrency exchange markets with top performances in the training periods would actually receive investment in the testing period. Furthermore, a portfolio management enables having control over how much capital is allocated to each cryptocurrency pair. On the whole, through an effective portfolio management, losses in unpromising markets are more likely to be avoided.

4.4.3   Case Study C—Time Rearrangement

From this case study onwards, the main objective is to describe the performance of each resampling method in detail. With this objective in mind, each case study contains statistics for a pair that yielded top results and another that yielded bottom results. Meanwhile, all results will be interpreted.

Firstly, a temporal graph showing the evolution of the average ROI per market for each trading signal is represented in Fig. 4.5. In other words, this figure, contains for each time instant, the respective average ROI of all 100 analysed markets. In this figure it can be observed that from May until September 2018 the analysed

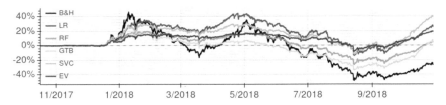

Fig. 4.5 Average accumulated ROI [%] for each instant in the test dataset with time resampled data

Table 4.2 Average obtained results for the Buy & Hold and each of the five methodologies employed with time resampled data

Parameter	B&H	LR	RF	GTB	SVC	Ensemble voting
Average obtained results (all markets are considered)						
Final ROI	−27.93%	25.61%	3.92%	3.30%	**39.53%**	18.72%
Accuracy	37.40%	54.77%	55.58%	54.84%	54.70%	**59.26%**
Negative log-loss	−21.6	−0.6931	−0.6896	−0.6890	−0.6757	**−0.6746**
Periods in market	100%	51.40%	44.38%	48.93%	44.94%	**27.73%**
Profitable positions	15.00%	**57.60%**	50.77%	53.36%	55.99%	54.95%
Profit per position	−27.93%	0.05%	0.01%	−0.02%	**0.09%**	0.06%
Largest gain	20.8%	18.10%	**20.76%**	19.63%	18.41%	15.32%
Largest loss	−48.7%	−15.20%	−14.51%	−14.94%	−16.18%	**−12.59%**
Max drawdown	77.0%	55.6%	56.8%	59.6%	52.5%	**42.5%**
Annual sharpe ratio	−0.352	0.075	−0.135	−0.327	0.228	**0.288**
Annual sortino ratio	−0.137	0.608	0.330	−0.074	0.985	**1.102**

market's price drops considerably, as can be confirmed by B&H's signal. EV and SVC were the methodologies that suffered the smallest losses in this period. Overall, most methodologies are quite conservative when the market suddenly rises, hence a B&H strategy earns more in these periods. On the other hand, this conservative behaviour is also responsible for minimizing losses in sudden price drops, contrarily to the B&H strategy.

Secondly, Table 4.2 contains the general statistics obtained for the time resampling method.

In this table, it is not easy to define a clear better method. Relatively to EV, SVC achieved a worse predictive power and is slightly more risky, yet obtained over twice

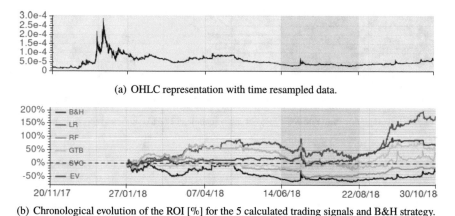

(a) OHLC representation with time resampled data.

(b) Chronological evolution of the ROI [%] for the 5 calculated trading signals and B&H strategy.

Fig. 4.6 ROI variations for currency pair POEETH (Po.et/Ethereum) with time resampling applied

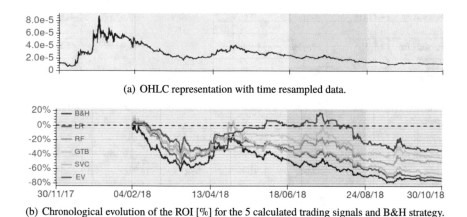

(a) OHLC representation with time resampled data.

(b) Chronological evolution of the ROI [%] for the 5 calculated trading signals and B&H strategy.

Fig. 4.7 ROI variations for currency pair ADABTC (Cardano/Bitcoin) with time resampling applied

the profits. Nonetheless, EV stands out due to the best accuracies and NLL as well as due to the remarkably small percentage of periods in market and top values in the risk metrics, thus suggesting that this is the least risky alternative. Thirdly, two specific markets and their return on investment evolution are represented in Figs. 4.6 and 4.7 and their statistics are shown in Table 4.3. In these figures, the top subfigure contains a candlestick representation of the utilized resampled data for each cryptocurrency pair. As can be seen, the background is divided into 5 different coloured periods. Each coloured period represents a different sub-dataset of the train-test splitting procedure, as was explained in Sect. 3.4.3. This splitting procedure divides the data in five intervals with the same amount of samples. Peculiarly in this case study, because the original data was uniformly resampled according to time, the output of

Table 4.3 Comparison between the Buy & Hold strategy and each of the five methodologies employed for the time resampling in the POEETH and ADABTC markets

Parameter	B&H	LR	RF	GTB	SVC	Ensemble voting
POEETH market with time resampling						
Final ROI	−28.01%	**179.9%**	−15.59%	51.76%	39.65%	71.26%
Accuracy	39.02%	55.51%	58.75%	58.36%	50.99%	**60.06%**
Negative log-loss	−21.06	−0.6851	−0.6732	−0.6840	**−0.6660**	−0.6730
Periods in market	100%	60.87%	35.21%	41.43%	73.30%	**31.03%**
Profitable positions	0%	**63.47%**	52.22%	54.35%	55.32%	55.67%
Profit per position	−28.01%	**0.205%**	−0.022%	0.060%	0.048%	0.102%
Largest gain	NA	3.55%	21.3%	16.67%	**25.87%**	16.67%
Largest loss	−28.01%	−20.62%	−20.30%	**−7.70%**	−20.98%	−20.5%
Max drawdown	70.5%	27.14%	63.51%	**19.16%**	57.27%	26.93%
Annual sharpe ratio	0.222	**2.190**	−0.057	1.174	0.900	1.471
Annual sortino ratio	0.368	**3.405**	−0.152	1.869	1.383	2.130
ADABTC market with time resampling						
Final ROI	−76.33%	−72.57%	−50.27%	−44.42%	−67.08%	**−34.07%**
Accuracy	35.54%	45.31%	55.21%	52.78%	43.41%	**57.22%**
Negative log-loss	−22.26	−0.7053	−0.6905	−0.6925	**−0.6632**	−0.6826
Periods in market	100%	79.95%	50.70%	54.15%	84.70%	**44.5%**
Profitable positions	0%	48.27%	44.80%	39.07%	49.08%	**51.17%**
Profit per position	−76.33%	−0.168%	−0.134%	−0.129%	−0.206%	**−0.080%**
Largest gain	NA	8.12%	10.77%	**21.75%**	19.41%	10.77%
Largest loss	−76.33%	−20.9%	**−14.36%**	−20.35%	−15.39%	−15.19%
Max drawdown	79.30%	74.20%	57.61%	53.87%	71.34%	**45.58%**
Annual sharpe ratio	−2.057	−2.350	−1.541	−1.051	−1.842	**−0.872**
Annual sortino ratio	−3.005	−3.372	−2.370	−1.719	−2.743	**−1.380**

this procedure yields five datasets with the exact same duration. Additionally, it is also visible that the ROI line only begins at the start of the second dataset, which is in accordance with the fact that the first sub-dataset is only used during the training procedure. In the following case studies figures with this same structure will be presented.

Note that, for the trading signals where stop-loss orders are active (all except the ones originated by B&H) throughout this whole chapter, the largest loss peaks at approximately –20%, the stop-loss activation percentage. Whenever the largest loss exceeds this value, a stop-loss order most likely had to be employed at least once by the system. Figure 4.6 contains the example of a market where this system performed remarkably well. From this figure and the upper half of Table 4.3 it is clear that the trading signal from the LR outperformed the remaining signals. EV obtained a clear second place with mostly above average results. Figure 4.7, on the other hand, contains a market with one of the worst performances obtained with this system. From this figure and the lower half of Table 4.3 it can be concluded that EV yielded the best results with lowest risk. However, if the trading signals from which it is derived from do not perform well, there is only so much EV can do. In any case, the B&H strategy was outperformed by the remaining trading signals as expected.

In conclusion, taking all figures and tables of this case study into account, the developed strategy is not flawlessly suited for this type of resampling. Neither B&H nor any of the five remaining models achieved absolutely formidable returns. In any case, it may be concluded the EV methodology is the superior one. It's worth praising EV's high predictive performances as can be perceived by the high accuracies and NLLs, implying that this system has predictive potential for this resampling method. This suggests that if this same machine learning formulae is combined with a more fine-tuned strategy, there is potential to achieve more impressive results, risk and return-wise.

4.4.4 Case Study D—Log Amount Rearrangement

This case study contains an analysis over the results obtained when the system utilizes logarithmic amount resampled data. Despite performing better than the time resampling method for the most part, it clearly underachieves the results of the two posteriorly analysed resampling methods. This case study follows the same structure of the previous one. Firstly, a temporal graph showing the instantaneous average ROI of the 100 analysed markets for each of the methodologies is shown in Fig. 4.8. In this figure, a drop also observed in the previous resampling method, is still visible from around May until September 2018. Nevertheless, this resampling method generates considerably higher ROIs relatively to time resampling. It is also observed that the B&H's ROI signal was clearly surpassed by all remaining trading signals in terms of profits.

Secondly, a table containing the general statistics obtained for this resampling method follows in Table 4.4. In this table, once more, the EV outperforms the remain-

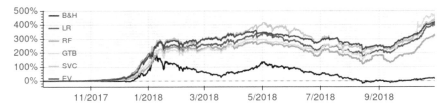

Fig. 4.8 Average accumulated ROI [%] for each instant in the test dataset with logarithmic amount resampled data

Table 4.4 Average obtained results for the Buy & Hold and each of the five methodologies employed with logarithmic amount resampled data

Parameter	B&H	LR	RF	GTB	SVC	Ensemble voting
Average obtained results (all markets are considered)						
Final ROI	16.61%	397.4%	325.8%	**431.4%**	431.2%	417.8%
Accuracy	41.31%	52.96%	52.77%	52.92%	52.59%	**55.02%**
Negative log-loss	−20.3	−0.7059	−0.7032	−0.6935	−0.7088	**−0.6866**
Periods in market	100%	58.21%	56.20%	57.66%	53.37%	**44.05%**
Profitable positions	25.00%	**61.40%**	54.46%	59.61%	59.97%	58.45%
Profit per position	**16.61%**	0.467%	0.350%	0.440%	0.511%	0.493%
Largest gain	**56.58%**	13.35%	17.45%	14.28%	16.94%	14.71%
Largest loss	−39.96%	**−12.86%**	−15.02%	−14.38%	−13.54%	−13.22%
Max drawdown	80.5%	57.3%	61.8%	62.6%	55.4%	**50.8%**
Annual sharpe ratio	0.070	1.173	0.643	0.638	1.018	**1.181**
Annual sortino ratio	0.618	**3.481**	2.006	2.267	3.042	3.293

ing methodologies. Even though on average the returns are slightly inferior to GTB and SVC, this method obtains the highest predictive power and the lowest risk out of all other trading signals. Anyhow, the B&H strategy clearly obtained the worst performance, as expected.

Thirdly, the results for two specific markets and their return on investment evolution are represented in Figs. 4.9 and 4.10 and their statistics are shown in Table 4.5. With the purpose of enabling comparison, these figures follow the same structure and contain the same markets as the figures presented in the previous case study. However, contrarily to the previous case study, note that with this, and the following

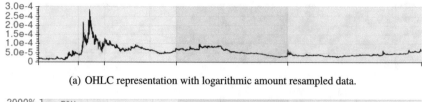

(a) OHLC representation with logarithmic amount resampled data.

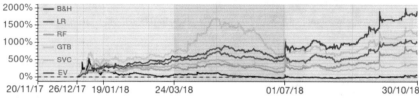

(b) Chronological evolution of the ROI [%] for the 5 calculated trading signals and B&H strategy.

Fig. 4.9 ROI variations for currency pair POEETH (Po.et/Ethereum) with logarithmic amount resampling applied

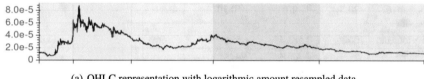

(a) OHLC representation with logarithmic amount resampled data.

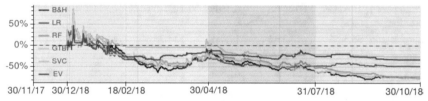

(b) Chronological evolution of the ROI [%] for the 5 calculated trading signals and B&H strategy.

Fig. 4.10 ROI variations for currency pair ADABTC (Cardano/Bitcoin) with logarithmic amount resampling applied

resampling methods, the different sub-dataset intervals, from the train-test splitting procedure, visibly, no longer contain the same duration.

Figure 4.9 contains the example of a market where this system performed well. From this figure and Table 4.5 it can be concluded that EV has the best predictive power and lowest risk. The remaining trading signals yielded results as good as expected except for the SVC who clearly underachieved despite on average yielding relatively better results as was seen in Table 4.4. Figure 4.10, on the other hand, contains the same market from the previous case study who underperformed. From this figure and Table 4.5, it can be concluded that SVC in an outlier fashion performed the best. Particularly RF and XGB generated awful results similar to the B&H strategy and LR is not particularly good as well. Hence, there was not much the EV method

Table 4.5 Comparison between the Buy & Hold strategy and each of the five methodologies employed for the logarithmic amount resampling in POEETH and ADABTC markets

Parameter	B&H	LR	RF	GTB	SVC	Ensemble voting
POEETH market with logarithmic amount resampling						
Final ROI	64.11%	1006.4%	693.9%	1241.7%	357.3%	**1901.5%**
Accuracy	43.25%	55.48%	52.66%	54.31%	53.29%	**55.90%**
Negative log-loss	−19.60	**−0.6862**	−0.7067	−0.6897	−0.6947	−0.6877
Periods in market	100%	48.4%	61.6%	52.4%	44.9%	**40.9%**
Profitable positions	100%	61.2%	55.4%	59.4%	56.0%	**61.4%**
Profit per position	**64.11%**	1.06%	0.75%	1.41%	0.66%	2.28%
Largest gain	**64.11%**	4.65%	0.59%	0.98%	1.87%	1.09%
Largest loss	NA	**−0.06%**	−8.12%	−2.95%	−8.12%	−2.25%
Max drawdown	90.10%	40.51%	27.40%	61.49%	60.15%	**18.26%**
Annual sharpe ratio	1.036	3.202	2.317	2.763	2.126	**4.346**
Annual sortino ratio	2.475	6.361	5.437	6.423	4.576	**9.753**
ADABTC market with logarithmic amount resampling						
Final ROI	−76.39%	−47.86%	−72.26%	−75.57%	**−24.97%**	−32.42%
Accuracy	40.54%	52.54%	49.12%	46.66%	**57.75%**	53.95%
Negative log-loss	−20.54	−0.6860	−0.7170	−0.7063	**−0.6779**	−0.6891
Periods in market	100%	47.6%	63.1%	71.7%	**14.3%**	33.7%
Profitable positions	0%	52.5%	40.59%	41.7%	**56.4%**	45.6%
Profit per position	−76.39%	**−0.13%**	−0.19%	−0.28%	−0.15%	−0.14%
Largest gain	NA	13.5%	29.2%	10.4%	17.5%	**33.4%**
Largest loss	−76.39%	−20.5%	−22.1%	−21.7%	**−8.41%**	−22.1%
Max drawdown	88.61%	73.15%	81.9%	87.0%	**39.0%**	55.4%
Annual sharpe ratio	−1.483	−1.430	−1.968	−1.664	−0.778	**−0.658**
Annual sortino ratio	−2.336	−2.053	−2.921	−2.531	−1.415	**−1.126**

could do. After all, this method is derived from the four trading signals received from the machine learning algorithm. For a given market, if the four trading signals outputted from the individual learning algorithms are unacceptable, most likely so will the EV signal be.

All in all, this resampling method successfully generates on average a much higher ROI and with lower risk relatively to the time resampling method as expected. Nevertheless, its predictive power is clearly inferior. This issue is discussed in further detail in Sect. 4.5.

4.4.5   Case Study E—Percentage Rearrangement

This case study contains an analysis over the results obtained when the system utilizes percentage resampled data. The EV results from this type of resampling, are only narrowly outperformed by amount resampling in terms of predictive power and final returns. Nevertheless, on average these results are significantly less risky. Due to its low risk and only slightly lower profits and predictive power, it is worth carrying out a more in depth analysis. This case study follows the same structure as the previous ones. Firstly, a temporal graph showing the instantaneous average ROI of the 100 analysed markets for each of the methodologies is presented in Fig. 4.11. In this figure, although not as significant as the previous resampling methods, the drop in the markets from May until September 2018 is also visible in this case. This resampling method generates much higher ROIs relatively to the logarithmic amount resampling, particularly for the SVC and EV methods. Once again, the signal from B&H was clearly surpassed by all remaining trading signals in terms of profits.

Secondly, a table containing the general statistics obtained for percentage resampling method follow in Table 4.6. In this table, even though, once again, on average EV's returns are slightly inferior to SVC, this method distinctly obtains the highest predictive power and the lowest risk out of all other trading signals, thus it may be concluded that the ensemble voting methods clearly is the top performer. The B&H strategy clearly obtained the worst average performance, as expected.

Thirdly, the results for two specific markets and their return on investment evolution are represented in Figs. 4.12 and 4.13 and their statistics are shown in Table 4.7.

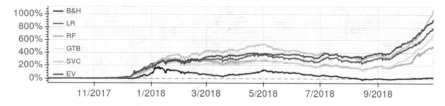

Fig. 4.11 Average accumulated ROI [%] for each instant in the test dataset with percentage resampled data

Table 4.6 Average obtained results for the Buy & Hold and each of the five methodologies employed with percentage resampled data

Parameter	B&H	LR	RF	GTB	SVC	Ensemble voting
Average obtained results (all markets are considered)						
Final ROI	16.41%	791.9%	494.4%	691.5%	**1062.9%**	922.74%
Accuracy	41.38%	53.03%	53.00%	53.06%	52.64%	**55.23%**
Negative log-loss	−20.2	−0.7112	−0.7028	−0.6928	−0.7187	**−0.6866**
Periods in market	100%	56.72%	55.80%	58.00%	52.30%	**43.28%**
Profitable positions	25.00%	**61.88%**	55.16%	57.96%	59.67%	59.04%
Profit per position	**16.41%**	0.82%	0.51%	0.67%	1.17%	0.97%
Largest gain	**56.27%**	20.86%	15.03%	16.29%	19.10%	13.59%
Largest loss	−39.86%	−13.23%	−14.18%	−14.23%	−13.85%	**−12.71%**
Max drawdown	80.5%	56.6%	59.9%	60.5%	55.4%	**50.1%**
Annual sharpe ratio	0.066	1.297	0.868	0.824	1.360	**1.496**
Annual sortino ratio	0.605	3.907	2.911	2.689	3.765	**4.325**

Once more, these figures follow the same structure and contain the same markets as the figures presented in the previous case studies.

Figure 4.12 contains the example of a market where this system performed well. From this figure and Table 4.7 it can be concluded that the signal from EV only obtained top results for the risk ratios. Results regarding ROI and specially predictive power are slightly inferior to the top values. In this example the top values are dispersed throughout the different trading signals, which goes to show that when utilizing varied learning algorithms, one's weakness can be overcome by another learning algorithm and, ultimately only the best traits of these algorithms are hopefully noticeable in the EV, which in this case, did not happen. Figure 4.13, on the other hand, contains the same market who underperformed from the previous case studies. From this figure and Table 4.7 it can be concluded that EV obtained top predictive results with a low risk, and even achieved a positive ROI, contrarily to the remaining trading signals. In fact, almost the opposite to what happened in the previous market is verified here. The trading signal from EV outdid most metrics of the remaining trading signals. This occurrence does not consistently happen, but is one potentiality of the EV. Obviously the contrary, where the trading signal produced by EV is the worst out of all, does happen as well, but it is a much rarer occurrence in this sys-

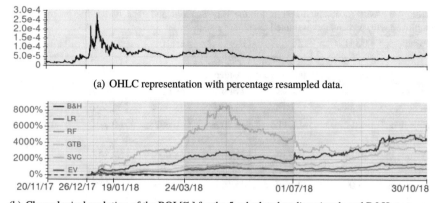

(a) OHLC representation with percentage resampled data.

(b) Chronological evolution of the ROI [%] for the 5 calculated trading signals and B&H strategy.

Fig. 4.12 ROI variations for currency pair POEETH (Po.et/Ethereum) with percentage resampling applied

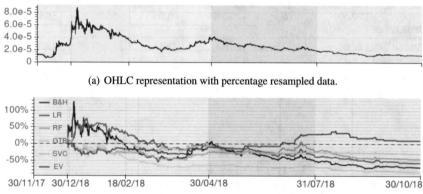

(a) OHLC representation with percentage resampled data.

(b) Chronological evolution of the ROI [%] for the 5 calculated trading signals and B&H strategy.

Fig. 4.13 ROI variations for currency pair ADABTC (Cardano/Bitcoin) with percentage resampling applied

tem as can be confirmed by the tables containing the average overall results for any resampling method, where the EV trading signal on average excels the remaining.

By now, it is evident that this resampling method is able to generate on average much higher ROIs than the previous two case studies. However, in order to achieve a more thorough analysis of this work's system, specific entry and exit points from periods with different characteristics are represented in Figs. 4.14 and 4.15. Figure 4.14 contains a volatile period of the cryptocurrency pair TRXBTC. Here it is visible that the system is mostly inside the market as the overall trend is bullish. Figure 4.15, on the other hand, contains a relatively calm period of the pair IOTABTC. The system is outside the market for days. By comparing the two figures, it is clear that the system

Table 4.7 Comparison between the Buy & Hold strategy and each of the five methodologies employed for the percentage resampling in POEETH and ADABTC markets

Parameter	B&H	LR	RF	GTB	SVC	Ensemble voting
POEETH market with percentage resampling						
Final ROI	64.24%	1436.3%	849.8%	3145.4%	**5747.3%**	4415.72%
Accuracy	42.59%	56.60%	52.33%	**56.71%**	53.92%	56.54%
Negative log-loss	−19.83	−0.6860	−0.7058	−0.6894	**−0.6824**	−0.6859
Periods in market	100%	46.5%	64.4%	**43.8%**	59.5%	44.87%
Profitable positions	100%	60.9%	54.1%	61.0%	**62.3%**	61.3%
Profit per position	**64.24%**	1.6%	0.96%	3.7%	2.1%	4.9%
Largest gain	**64.24%**	2.61%	2.50%	4.76%	6.4%	7.3%
Largest loss	NA	**−2.2%**	−8.9%	−3.7%	−8.2%	−4.3%
Max drawdown	89.98%	**16.60%**	54.60%	20.03%	70.94%	23.92%
Annual sharpe ratio	1.041	3.969	2.661	3.868	3.404	**3.970**
Annual sortino ratio	2.510	9.326	6.125	10.192	10.829	**11.927**
ADABTC market with percentage resampling						
Final ROI	−71.06%	−56.81%	−45.75%	−85.70%	−29.08%	**10.12%**
Accuracy	40.0%	52.04%	54.88%	50.24%	55.97%	**59.02%**
Negative log-loss	−20.72	−0.6921	−0.6903	−0.6941	**−0.674**	−0.6837
Periods in market	100%	58.4%	52.5%	61.9%	**23.6%**	29.4%
Profitable positions	0%	**58.6%**	47.7%	48.4%	47.1%	53.8%
Profit per position	−17.06%	−0.14%	−0.082%	−0.20%	−0.19%	**0.026%**
Largest gain	NA	9.5%	9.4%	**14.4%**	12.5%	9.2%
Largest loss	−71.06%	−20.3%	−11.9%	−24.2%	**−8.8%**	−11.8%
Max drawdown	88.64%	75.77%	57.4%	87.7%	**32.7%**	35.7%
Annual sharpe ratio	−1.483	−1.190	−1.028	−3.168	−1.013	**0.471**
Annual sortino ratio	−2.303	−1.812	−1.637	−4.061	−1.682	**0.662**

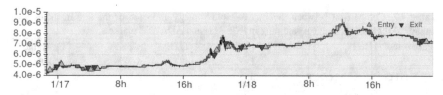

Fig. 4.14 Entry and exit points for the TRXBTC (Tron/Bitcoin) pair during 17 and 18 January 2018 for percentage resampling

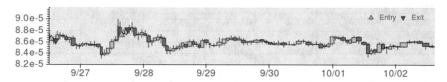

Fig. 4.15 Entry and exit points for the IOTABTC (Internet of Things Application/Bitcoin) pair from mid September 26 until mid October 2 2018 for percentage resampling

is more inactive during calm intervals, which is in accordance with the intention of taking advantage of highly volatile situations.

In conclusion, this resampling procedure clearly outperforms the logarithmic amount procedure. Hence, it is relatively preferable, namely because, as was mentioned in Sect. 3.2, these two procedures have similar properties. The fact that these two procedures generate the sub-datasets intervals with the closest separation dates, and consequently similar durations for the respective intervals, supports their similarity. Relatively to the time procedure, even though the profits were clearly superior, likewise the logarithmic amount, its predictive power is clearly inferior. This last issue is discussed in further detail in Sect. 4.5.

4.4.6 Case Study F—Amount Rearrangement

This case study contains an analysis over the results obtained when the system utilizes amount resampled data. This type of resampling procedure generates on average the best performance in terms of the ROI metric for the trading signals generated by EV. Hence, similarly to the previous case study, a more in depth analysis will be carried out. For continuity's sake and to enable a more efficient comparison, this case study follows the same structure as the previous ones. Firstly, a temporal graph showing the instantaneous average ROI of the 100 analysed markets for each of the methodologies is represented in Fig. 4.16. In this figure, similarly to percentage resampling, a slight drop is still visible from around May until September 2018. This resampling method generates an overall higher ROI relatively to the percentage resampling method. Once again, the B&H signal was clearly surpassed by all remaining trading signals in terms of returns.

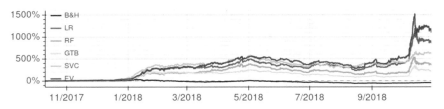

Fig. 4.16 Average accumulated ROI [%] for each instant in the test dataset with amount resampled data

Table 4.8 Average obtained results for the Buy & Hold and each of the five methodologies employed with amount resampled data

Parameter	B&H	LR	RF	GTB	SVC	Ensemble voting
Average obtained results (all markets are considered)						
Final ROI	−46.92%	859.1%	354.2%	215.5%	617.8%	**1100.3%**
Accuracy	40.84%	53.24%	53.12%	53.20%	53.58%	**55.61%**
Negative log-loss	−20.4	−0.7021	−0.7012	−0.6921	−0.6868	**−0.6839**
Periods in market	100%	57.75%	55.28%	55.56%	50.96%	**44.41%**
Profitable positions	13.00%	**59.72%**	52.80%	55.49%	57.42%	57.10%
Profit per position	−46.92%	0.919%	0.357%	0.210%	0.671%	**1.22%**
Largest gain	8.66%	18.26%	16.08%	**18.94%**	19.12%	16.44%
Largest loss	−55.6%	−16.28%	−16.31%	−17.00%	−14.48%	**−14.10%**
Max drawdown	81.6%	61.0%	63.7%	65.3%	55.5%	**53.7%**
Annual sharpe ratio	−0.440	0.531	0.276	0.114	0.786	**0.816**
Annual sortino ratio	−0.411	1.500	1.410	0.744	2.479	**2.565**

Secondly, a table containing the general statistics obtained for amount resampling method follow in Table 4.8. In this table, yet again, the trading signal produced by EV clearly outperforms the remaining methodologies. This method obtains the highest profits, the best predictive power and the lowest risk out of all available trading signals. The B&H strategy visibly performed the worst as usual. Relatively to the previous case studies, it may be concluded that on average, this resampling method generates the highest profits, but is riskier than both logarithmic amount and percentage resampling.

Thirdly, the results for two specific markets are represented in Figs. 4.17 and 4.18 and their statistics are shown in Table 4.9. Likewise, these figures follow the same

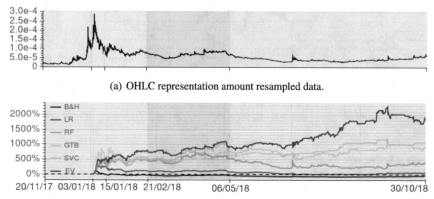

(a) OHLC representation amount resampled data.

(b) Chronological evolution of the ROI [%] for the 5 calculated trading signals and B&H strategy.

Fig. 4.17 ROI variations for currency pair POEETH (Po.et/Ethereum) with amount resampling applied

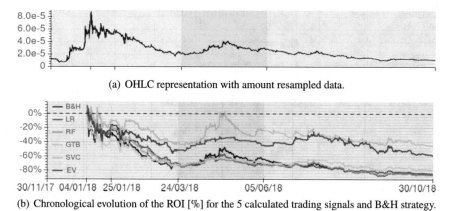

(a) OHLC representation with amount resampled data.

(b) Chronological evolution of the ROI [%] for the 5 calculated trading signals and B&H strategy.

Fig. 4.18 ROI variations for currency pair ADABTC (Cardano/Bitcoin) with amount resampling applied

structure and contain the same markets as the figures presented in the previous case studies. Figure 4.9 and Table 4.9 contain the example of a market where this system performed notably well. Once more, the trading signal from the EV overcomes in all aspects the remaining signals. Nonetheless, it is observable that in this resampling method worse results were obtained for this market relatively to when percentage resampled data was used. The fact that lower values were obtained can be up to some degree blamed on the lower predictive performance and the higher riskiness, as can be seen by comparing the metrics from this table with Table 4.7. Figure 4.10, on the other hand, contains the market who underperformed. From these figures and their respective values in Table 4.9, a conclusion similar to the one taken in the previous case study may be taken. The SVC's signal performance, one more time, exceeded

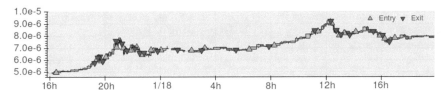

Fig. 4.19 Entry and exit points for the TRXBTC (Tron/Bitcoin) pair during most of 17 and 18 January 2018 for amount resampling

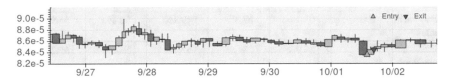

Fig. 4.20 Entry and exit points for the IOTABTC (Internet of Things Application/Bitcoin) pair from mid September 26 until mid October 2 2018 for amount resampling

the remaining trading signals. However, in this case, similarly to the logarithmic amount case study, the EV was unable of excelling both SVC and the remaining trading signals.

For comparison purposes, the entry and exit points of the same periods and markets used in the previous case study, but with data resampled according to an amount, are represented in Figs. 4.19 and 4.20. Figure 4.19 contains a highly volatile period of the cryptocurrency pair TRXBTC. Relatively to the previous case study (Fig. 4.14), it is clear that with amount resampled data, the trading signal typically does not keep the same long position active for as much time. With this type of resampling, relatively to percentage resampling, the triggered long positions are of shorter durations. Therefore, this strategy is prone to sustain heavier losses, namely due to transfer fees, which explains the larger risk verified. Figure 4.20, on the other hand, contains a relatively calm period of cryptocurrency pair IOTABTC. Note that all candles have durations of above 2 h due to the reduced price variation. The largest variation seen in this figure is of around 5% in the course of 12 h, this same pair earlier in 2018 on a regular basis had this same percentual variation in a matter of minutes. Once again, the judgement that this system is clearly more active in volatile moments is upheld.

In conclusion, this resampling procedure outperforms the results obtained with logarithmic amount resampled data in all aspects. Regarding, percentage resampled data, generally this resampling procedure yields a better ROI and a slightly superior predictive power, yet this procedure clearly has a higher risk associated. Once again, this resampling method obtains clearly superior ROI's relatively to the time procedure, but the predictive power is certainly inferior, likewise the amount and logarithmic amount procedures. This issue is discussed in further detail in Sect. 4.5.

Table 4.9 Comparison between the Buy & Hold strategy and each of the five methodologies employed for the amount resampling in POEETH and ADABTC markets

Parameter	B&H	LR	RF	GTB	SVC	Ensemble voting
POEETH market with amount resampling						
Final ROI	−39.35%	67.34%	367.7%	904.0%	1070.0%	**1930.8%**
Accuracy	42.31%	52.96%	52.41%	56.33%	54.90%	**57.18%**
Negative log-loss	−19.92	−0.6881	−0.6970	−0.6870	**−0.6793**	−0.6817
Periods in market	100%	63.4%	67.3%	49.4%	52.9%	**45.9%**
Profitable positions	0%	56.9%	53.4%	57.7%	58.6%	**59.2%**
Profit per position	−39.35%	0.060%	0.39%	0.98%	1.2%	2.0%
Largest gain	NA	4.11%	2.4%	4.1%	3.7%	**11.6%**
Largest loss	−39.35%	−16.7%	−7.7%	−0.84%	**−0.66%**	−1.7%
Max drawdown	90.3%	75.3%	65.0%	37.1%	40.4%	**24.9%**
Annual sharpe ratio	0.2394	0.9534	1.855	2.051	2.799	**3.092**
Annual sortino ratio	0.4224	1.861	4.854	7.718	7.222	**11.979**
ADABTC market with amount resampling						
Final ROI	−85.87%	−84.06%	−86.45%	−79.24%	**−45.78%**	−58.85%
Accuracy	39.37%	51.28%	49.51%	51.23%	52.96%	**54.66%**
Negative log-loss	−20.94	−0.6916	−0.7176	−0.6935	**−0.6732**	−0.6882
Periods in market	100%	70.0%	71.1%	60.9%	64.3%	**59.6%**
Profitable positions	0%	**52.1%**	46.6%	49.3%	50.4%	50.7%
Profit per position	−85.87%	−0.135%	−0.213%	−0.200%	−0.084%	**−0.113%**
Largest gain	NA	12.2%	13.5%	10.2%	**16.4%**	15.4%
Largest loss	−85.87%	−12.9%	−21.6%	−11.5%	**−8.4%**	−9.7%
Max drawdown	88.58%	85.5%	87.5%	79.9%	**50.6%**	61.9%
Annual sharpe ratio	−2.224	−2.802	−3.018	−2.475	**−0.960**	−1.394
Annual sortino ratio	−3.197	−3.845	−4.025	−3.341	**−1.645**	−2.015

4.5 Conclusions

In this chapter, firstly the metrics utilized for analysis and comparison of the different trading signals and resampling methods employed in this system are described, as well as a simple investment strategy to be used as baseline. Posteriorly, particularly noteworthy case studies are analysed.

Regarding the analysis done to case study A, Sect. 4.4.1, it's worth adding that, to some degree, this hypothesis may be extended to the remaining markets. The markets utilized may not be able to accept a trade instantaneously as the algorithm intends, in a real case scenario this situation of illiquidity could take a toll on the system's performance, but as mentioned in Sect. 3.5, this is a limitation of a backtest trading simulation.

The second case study, Sect. 4.4.2, reveals that the ensemble voting classifier matches the initial expectations. The great majority of times the trading signals from this forecasting method are the most profitable and typically achieve top performance according to all metrics used for evaluation.

The succeeding case studies, Sect. 4.4.3 through Sect. 4.4.6, individually analyse the trading performance of this work's system for different resampling methods. In terms of final profit, all performed better than the ordinarily utilized time resampling method presented in Sect. 4.4.3. In any of these four case studies, the trading signals originated by EV customarily are the less risky and contain the best predictive power. Even though only on amount resampling did EV on average obtain the top returns, in the remaining resampling procedures, the ROI obtained by EV is well above average, nearing the top value for this metric out of all methods. Despite logarithmic amount resampling yielding on average a higher ROI than time resampling, according to this same metric, it was clearly inferior to the amount and percentage resampling procedures. Relatively to percentage resampling, amount resampling obtains on average a slightly superior ROI, but this fact could be attributed to its higher risk. Furthermore, the difference in predictive power is almost negligible. Therefore, it is hard to pick a clear preferable choice between these two.

In any case, as a rule, all three alternative resampling procedures consistently provide a higher ROI relatively to time resampling, suggesting that financial time series resampled in accordance to an alternative metric, rather than time, have the potential of creating a signal more prone of earning larger profits. Anyhow, independently of the resampling method, this system is clearly more active in volatile periods than calm periods as was pretended. Throughout these last four case studies, it was verified that specific markets, such as ADABTC, will invariably yield losses independently of the resampling procedure. A portfolio management could possibly withdraw these markets from the list of tradeable markets in order to prevent these types of losses. On the upside, it was also verified that if this work's system was able of turning a profit with a time resampled financial series, most likely, the same given market attains far larger returns with any of the three alternative resampling procedures. One important takeaway worth noting is that, independently of the utilized strategy, higher accuracies or NLL do not inevitably translate into a higher ROI or a lower risk.

One last point must be addressed: the fact that, time rearrangement on average, holds the lowest ROI, but contradictorily also obtains the best accuracies and negative log-losses out of all the resampling methods employed. One possible explanation is because most investors utilize similar investment strategies and prediction algorithms based on similarly time sampled data. Resemblance in prediction algorithms imposes a positive feedback loop, where the resultant time sampled data is continuously being shaped. Because the algorithms and strategies utilized by investors are related and due to their collective influence, the current ongoing time series is constantly being heavily impacted. Hence, when an algorithm similar to the ones who collectively crafted and whose influence is deeply embedded in a given time series intends to backtest trade with this same time series, it seems plausible that its overall predictive power is enhanced. Consequently, the two metrics associated with machine learning turn out relatively better when the original time series is utilized, this is, when time resampling is applied. Time series sampled according to time obtained lower returns relatively to the alternative resampling procedures, possibly on account of the following two reasons: not having the properties mentioned in Sect. 3.2 that make a time series prone to larger earnings; this work's time sampling frequency may be out of phase relatively to the majority of investors. While it is common to utilize 1, 10 or 60 minutes as sampling frequency, in this work the utilized frequency is usually an integer anywhere between 1 and 10 min. As a result, the maximum possible value for the return on investment could be limited.

This hypothesis stems out of the fact that, nowadays human investors or investment bots, may coincidentally or purposefully (if a strategy or bot is intentionally teamed with others), artificially inflate or deflate prices and propagate feedback loops (an extreme example of this propagation, albeit usually unintentional, are flash-crashes[2]). Investors purposefully teaming up their strategies or bots is considered market manipulation. Many different schemes of market manipulation have been detected in the cryptocurrency exchange market world (such as wash trading, spoofing, pump-and-dump, etc.), but as of now, are tolerated due to the non-regulated nature of these markets. Nevertheless, market manipulation is a topic that falls out of the scope of this work and, in order to assign it as the unambiguous source of this issue would require further investigation.

References

1. Magdon-Ismail M, Atiya AF (2015) An analysis of the maximum drawdown risk measure. Citeseer
2. Mallqui DC, Fernandes RA (2019) Predicting the direction, maximum, minimum and closing prices of daily Bitcoin exchange rate using machine learning techniques. Appl Soft Comput 75:596–606

---

[2]Very rapid, deep and volatile fall in prices occurring within a short time period. Generally stem from high-frequency trading bots.

Chapter 5
Conclusions and Future Work

5.1 Conclusion

In this work a system combining several machine learning algorithms with the goal of maximizing predictive performance was described, analysed and compared to a B&H strategy. All aspects of this system, namely the target formulation, were designed with the objectives of maximizing returns and reducing risks always in mind. To validate the robustness of this system's performance, this system was tested on 100 different cryptocurrency exchange markets through backtest trading.

Based upon this work, it may be concluded that all four distinct learning algorithms consistently bared positive results, better than random choice or the B&H strategy. Nonetheless, the trading signal generated by the ensemble voting method by far provides the best results. Particularly with ensemble voting method when time resampling is employed, the predictive power of this work's proposed system is on average superior to the state-of-the art in forecasting the direction of prices with a sampling frequency in the order of minutes, as can be seen by comparing Sect. 2.3, to this work's case studies, Sect. 4.4. Therefore, the task of creating a successful forecasting system may be deemed as accomplished.

In addition, the outcome of utilizing several different resampling procedures was analysed. On average the three alternative resampling procedures studied in this work generate significantly higher returns than commonly utilized time sampled data. In terms of the returns on investment obtained, resampled data according to an amount was the top performer, followed by the percentage and lastly the logarithmic amount. While ordinary time resampling data originates the best accuracies and NLLs, the lowest returns out of all tested resampling methods were counterintuitively also obtained with this method. A possible reason for this occurrence is discussed in Sect. 4.5. Nevertheless, the overall results were in accordance with what was expected: rearranging financial time series according to an alternative metric does in fact construct a series prone to generating larger returns, which is in accordance with one of the main intentions of this work. It should be pointed out that throughout all

© The Author(s), under exclusive license to Springer Nature Switzerland AG 2021

T. A. Borges and R. Neves, *Financial Data Resampling for Machine Learning Based Trading*, SpringerBriefs in Computational Intelligence, https://doi.org/10.1007/978-3-030-68379-5_5

stages of this work, a primary priority was ensuring that each different resampling method was tested under equal and fair conditions in all modules of the system, so as to avoid any unfairness that could invalidate the final results.

It is worth noting that ROIs obtained in this system with the alternative procedures are in some markets excessive. This fact is attributed to the absence of bid-ask spread data, as was mentioned in Sect. 3.4.2. Putting it differently, this work shows that the three alternative methods offer more profitable time series relatively to time resampling, however, it is unlikely that this system would do as well, return-wise, in an actual real scenario in face of bid-ask spreads.

In conclusion, the initial objectives were accomplished and the obtained results are promising: several methods for predicting the direction of price in the cryptocurrency market were developed, particularly the EV, were successful and the alternative resampling procedures certainly have potential in generating larger returns relatively to the commonly put in practice time resampling procedures.

5.2 Future Work

Despite having accomplished this work's objectives, there is always room for improvement. For future work, besides perfecting the investment strategy that defines the entry and exit points, several limitations of this work as well as possible future directions or improvements are suggested:

- Firstly, the bid-ask spreads should be gathered and incorporated into this system.
- In this thesis the utilized resampling thresholds are constant. In future work, adjusting these values dynamically as a function of specific indicators could be explored. Additionally, the parameters of each technical indicator as well as of each individual classifier are also fixed. Either a method for finding the ideal values of each parameter such as a genetic algorithm could be introduced, or a more extensive grid searching procedure could be introduced.
- Regarding the set of features utilized as input, additional indicators could be experimented with, specifically fundamental indicators (such as market capitalization) or data extracted from social media. Even though the importance of these additional indicators is unclear in literature, it is plausible that adding these alternative indicators to a set of technical indicators, may give an extra advantage to the system's performance. Specifically because in the market under analysis, the cryptocurrency exchange market, schemes of market manipulation are allowed and disclosed through social networks.
- Mechanisms for portfolio management, money management[1] or both, could be put in place to avoid the riskiest and loss-making markets.
- In this work the financial data was pulled from Binance due to its accessible API as well as because it enables trading a set of cryptocurrency pairs that are nowhere

---

[1]Method for managing the size of the position taking the risk and total capital available into consideration.

else offered. In order to correctly evaluate this forecasting system, the creation of a faithful and realistic simulation was a main priority of this work, the conditions imposed by the exchange, namely being limited to taking only long positions were complied. If Binance were to enable short positions, the final results could potentially be improved. There are other exchanges who allow short positions, nonetheless, it is unlikely to find one that offers these remarkably volatile exotic cryptocurrency pairs. In these exchanges (such as in Bitfinex, eToro, etc.), most likely only the most popular cryptocurrencies are traded against the USD.

Printed in the United States
By Bookmasters